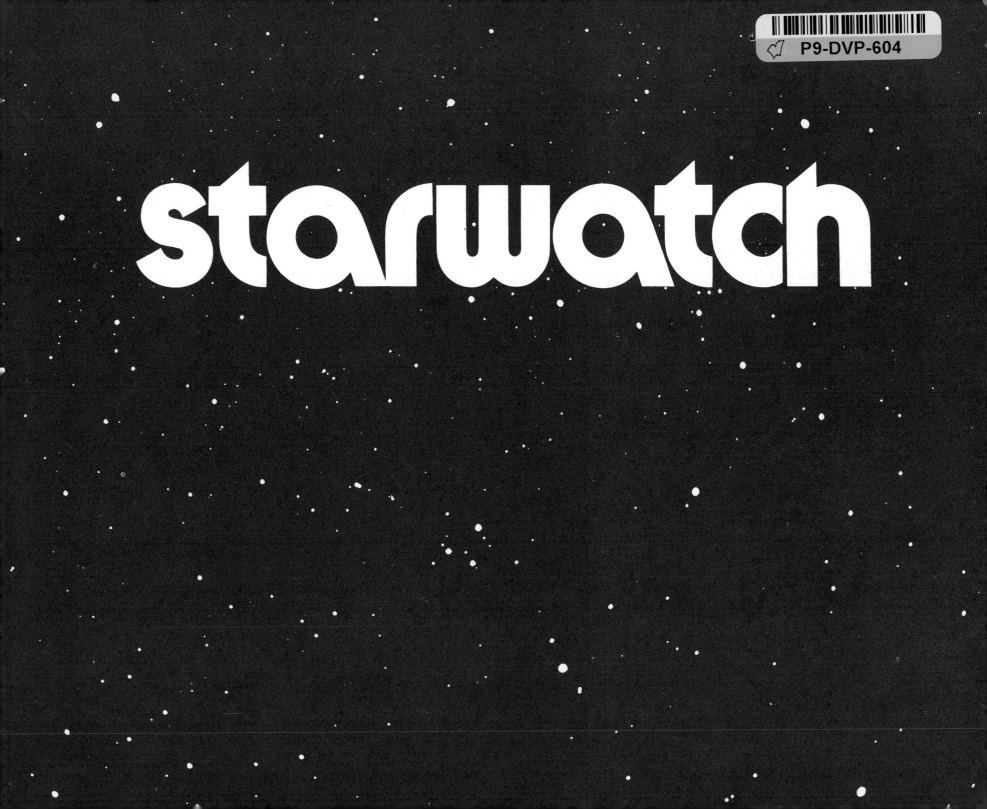

starwatch

This book is dedicated to the memory of Shawn Gregory.

The heavens declare the glory of God;
and the firmament showeth his handywork.

                    —PSALMS 19:1

"HOW LONG IS ETERNITY?"

Heinrich Schlieman, nineteenth-century genius,
discoverer of the ruins of the ancient city of Troy,
answered his son's question with a sweep of his hand:

"Picture a wall of marble ten yards high and ten yards wide
stretching all the way from here in Athens
to the little Greek harbor town of Piraeus, there in the distance.
Now imagine a bird coming along once every thousand years
to draw a silken scarf along the entire length of this gigantic slab of stone.
By the time friction will have worn the wall of marble down
to the thickness of a mere sliver,
one second of eternity will have passed."

starwatch

Ben Mayer A.R.C.A.

A PERIGEE BOOK

Perigee Books
are published by
The Putnam Publishing Group
200 Madison Avenue
New York, New York 10016

Back cover photos by Ben Mayer
Book design by Ben Mayer
Illustrations, star maps and astrophotography by Ben Mayer

Library of Congress Cataloging in Publication Data

Mayer, Ben.
  Starwatch .

  1. Astronomy—Amateurs' manuals. I. Title.
II. Title: Starwatch.
QB63.M445   1984   523   83-23820
ISBN 0-399-51008-7
ISBN 0-399-51009-5 (pbk.)

First Perigee printing, 1984
Printed in the United States of America
1 2 3 4 5 6 7 8 9

The author wishes to thank Dr. Hans Vehrenberg and
the Treugesell Verlag, Düsseldorf, for giving permission
to reproduce the copperplate illustrations from the
1782 Johann Bode star atlas.

Thanks are also due to Dr. Tom Kuiper and Dennis di Cicco
for proofreading the general chapters
and the constellation pages respectively.

The variable star material on pages 121-22 was reproduced
with special permission of the
American Association of Variable Star Observers.

STARFRAMES and PROBLICOM are trademarks of the author.

CONTENTS

HOW FAR IS A LIGHT-YEAR ?

In one second light travels a distance of	186,000 miles.
In one minute (60 seconds) light travels	11,160,000 miles.
In one hour (60 minutes) light travels	669,600,000 miles.
In one day (24 hours) light travels	16,070,400,000 miles.
In one year (365 days) light travels	5,865,696,000,000 miles.
One light-year equals about	6,000,000,000,000 miles.

A LIGHT-YEAR IS A UNIT OF LENGTH — NOT OF TIME.

RADIO-TELESCOPE, PARKES, AUSTRALIA

THE STARS AWAIT

You can set out at any time on your journey to the stars. Nothing will hinder you if you bring along that wonder, curiosity and imagination which the universe has always inspired in human hearts. Since earliest times, real and mystical relationships were sensed between the heavens and earth, connecting celestial events with planting and harvesting seasons. Ancient priests worshiped the sun, our nearest star. They studied the fiery disc and the constellations among which it seemed to travel. Biblical shepherds in Galilee viewed the skies as familiar friends.

Today we stand at the edge of space. Mirrored electronic eyes peer deep into the interstellar past. Huge, dish-shaped gossamer ears strain to hear signals from extraterrestrial intelligence. Comets, eclipses and fireballs that once caused terror now receive careful observation, and fear has given way to scientific enlightenment. We have learned much. Yet, infinitely more remains to be discovered.

This sky album shares some of the oldest and newest basic knowledge we have plucked from the heavens. It was designed by a lover of the stars, an amateur astronomer (*amator*=lover). Rather than try to offer the entire celestial sphere all at once, it opens windows on the twelve zodiacal starfields and twelve other constellations, all readily visible over the course of a year from northern latitudes.

A complete new concept of "starframes" is introduced here to help you find each constellation, even from city locations. Traced from pages of this book and held against the night sky, your starframes will allow you to match the brightest stars with their actual positions to draw you deeper into easy stargazing. You will soon find that there are dimensions in space that pass all understanding, and questions which no science book may ever be able to answer clearly. But the gratifying realization is that the most learned professor of astronomy and the humblest beginning stargazer know equally little about the mysteries of the universe.

Concerning equipment, your eyes will do just fine for a start. Binoculars will give you more reach than was available to famous scholars up until the time of Galileo. Almost any commercially available telescope will raise your view to that of royal astronomers in yesteryear's Greenwich Observatory in England.

The newest collector of starlight, an ordinary 35mm camera, will record more than the human eye could possibly see. Together with a "Problicom," a gadget you can build yourself at home (pages 132-33), photography will put you on the threshold of discovery.

Nature's most magnificent free show will always offer more than we can see even with the best telescopes, so you must draw on your most important resources: your mind's eye and your soul. In this area your equipment may exceed that of the world's leading scientists. Once you have been humbled by the realization that our entire solar system is an insignificant part of the Milky Way galaxy, that what appears to be a fuzzy star is really a ball of a million suns, your horizons will expand to new frontiers. Your first naked-eye glimpse of the Andromeda galaxy (pages 102-5), a member of our "local group of galaxies," will set your sights two million light-years farther out. There need be no limit to your vision. The awareness that there are thousands of millions more galaxies beyond ours may allow you to answer for yourself the ultimate question: Are we alone?

STONEHENGE, SALISBURY PLAIN, ENGLAND

THE ZODIAC

There is a zoo in the Zodiac. With the exception of Libra (the Scales), all of the traditional constellations of the Zodiac depict living things, people or animals. The word has a Greek ancestry. The "Band of the Zodiac" girdles our globe as shown on the opposite page. Running through the length of this ribbon in the sky, along its very middle, lies the "ecliptic," the path on which ancient astronomers believed the sun to be traveling around the earth. Today we know that it is the earth which moves around the sun, creating the appearance of a sun in motion. The twelve main starfields through which the sun appears to move are called the constellations of the Zodiac.

It takes our planet 365 day/night turns on its axis to complete one swing around the sun. This determines the length of our year. As the earth slowly turns on its axis, different parts of the sky become visible to us. The motions are best observed during the night, from the side of the earth then facing away from the sun. Changes in the sky cannot readily be seen during the day from the sunlit half of earth. Daylight makes the stars invisible except at dawn and dusk. As the earth moves slowly in its one-year orbit around the sun, we can actually experience the earth making one revolution in relation to the stars.

Because the earth is tilted, the sun is north of the equator for half of the year and south of it for the remaining half, resulting in changes of seasons (pages 140-41).

There are eighty-eight star groupings, called constellations, in the sky. They are evenly distributed in the celestial sphere, which can best be described as an imaginary transparent bubble surrounding our earth. The ancients thought that the stars were attached to the inside of this hollow globe, all at the same distance from the earth, which they believed to be at the center of the sphere. They divided the crystal bubble into twenty-four melon-peel slices (page 95), and numbered the lines dividing these slices zero through twenty-four. (The zero line and the twenty-fourth are one and the same.) To this day these lines are used to describe the east-west positions of stars. They established that zero was the place on the equator of the celestial sphere where the sun each year crosses from the southern to the northern celestial hemisphere. This zero point fell in the constellation of the Ram (Aries) and was called the "first point in Aries." The twenty-four-hour numbering system was called "right ascension" (R.A.) because of the order in which stars and constellations rose in the east. Thus the position of Gemini is at seven hours R.A., Leo at ten hours R.A. and the starfield Lyra, for example, at the nineteen-hours R.A. position.

Modern science has proved that the stars all lie at vastly different distances from our earth and that the sun, in fact, is just one of these stars. But astronomers still employ the convenient concept of the neatly divided celestial sphere invented by early scholars. Even today we continue to use the twenty-four hours of right ascension (with minutes and seconds for precision) to pinpoint the positions of all celestial objects. Locations in the east-west axis of the sky will always be referred to by R.A. hours. As we shall presently see, the north-south positions are established in vertical degrees of "declination." Together with the lines of right ascension, this gives us a tidy grid system with which to find our way in the sky.

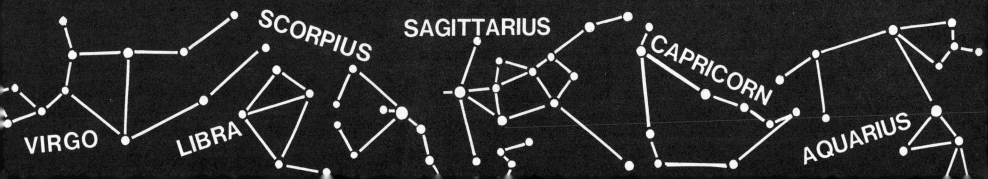

VIRGO   LIBRA   SCORPIUS   SAGITTARIUS   CAPRICORN   AQUARIUS

EARTH IN MOTION

In your mind, inflate the imaginary sphere of the ancients, the crystal ball bubble to which all constellations are attached. Like the earliest astronomers, we will divide the hollow ball into melon-peel slices numbered zero to twenty-four. Let us go by spacecraft to the north celestial pole (NCP) high within this celestial sphere. This will put us very close to Polaris, the North Star (see pages 12 and 36). The north pole of our earth will lie directly below us, as illustrated on the left, and we will be able to view the entire Northern Hemisphere of earth. Surrounding the equator of our planet but far out in space will be the celestial right ascension equator, divided into twenty-four hours, with the constellation R.A. positions marked on it as shown.

It is midnight on July 1 as we look down on an observer in the Northern Hemisphere who is viewing the skies through a small wire frame held at arm's length. We know that the observer is lying on the curved surface of the earth, but, since the planet is very large, four horizons present themselves to the observer, whose head is toward the north, feet pointing south. We will call the observer's position the "flatform." It is important to remember that the stars are essentially fixed in position on the celestial sphere. They do not move perceptibly from year to year, or even from century to century.

On July 1 at midnight, the constellation of Lyra is at its highest point in the sky. We say it culminates for midnight observers. Midnight is the opposite of noon when the sun is at its highest. Now consider the 365 fast *daily* turns of the earth which give us days and

shown in the diagram, Lyra *seems* to our observer to move west hour by hour until eventually it "sets." The observer who has viewed Lyra at midnight on July would have to look a little to the west to observe it an hour later (at 1 A.M.).

Observations on successive weeks will soon show the additional slow *annual* revolution of the earth in relation to the stars. We notice that Lyra culminates a midnight only at the beginning of July. On July 15 Lyra will *seem* to have moved to the right (westward) Our earth's slow motion around the sun will have placed the flatform observer beneath the next melon slice where Lyra's neighbor Cygnus is located

The observer on the flatform can compensate for the fast daily (or nightly) turns and the slow annual revolution of the earth by holding the starframe in different positions, as shown, depending upon the hour of the night or the day of the month. This book is arranged to show constellations as they culminate at midnight This does not mean that you have to do your observing at that hour. As you can see at the left, the culminating position moves with time. Thus, Lyra culminates on the July 1 at midnight, on July 15 at 11 P.M. on August 1 at 10 P.M. and on August 15 at 9 P.M.

Do not be disappointed if words alone do not, at first completely clarify the constant changes in the sky. I took humankind thousands of years, from the darkes beginnings of civilization until fairly recently, to fathom the magnificent order of the heavens. Careful observations should soon give insight into the motion of stars. At least we know today that we will not fall of our world if we step too close to the edge of ou

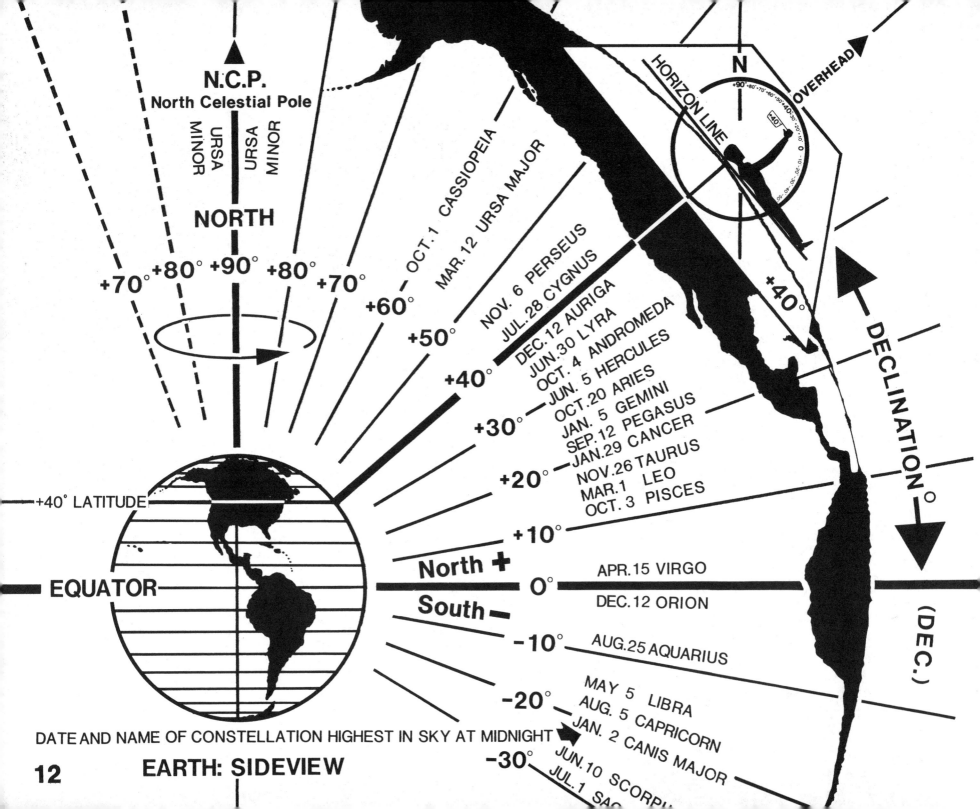

N.C.P.
North Celestial Pole

URSA MINOR

URSA MINOR

NORTH

+70° +80° +90° +80° +70° +60°

+50°

+40°

+30°

+20°

+10°

+40° LATITUDE

EQUATOR

North +

South −

0°

−10°

−20°

−30°

OCT. 1 CASSIOPEIA
MAR. 12 URSA MAJOR
NOV. 6 PERSEUS
JUL. 28 CYGNUS
DEC. 12 AURIGA
JUN. 30 LYRA
OCT. 4 ANDROMEDA
JUN. 5 HERCULES
OCT. 20 ARIES
JAN. 5 GEMINI
SEP. 12 PEGASUS
JAN. 29 CANCER
NOV. 26 TAURUS
MAR. 1 LEO
OCT. 3 PISCES

APR. 15 VIRGO
DEC. 12 ORION

AUG. 25 AQUARIUS

MAY 5 LIBRA
AUG. 5 CAPRICORN
JAN. 2 CANIS MAJOR
JUN. 10 SCORPIUS
JUL. 1 SAG

HORIZON LINE

N

OVERHEAD

+40°

DECLINATION° (DEC.)

DATE AND NAME OF CONSTELLATION HIGHEST IN SKY AT MIDNIGHT

12   **EARTH: SIDEVIEW**

WHAT'S UP?

What is "up" for Northern Hemisphere observers is "down" for those in the southern part of our globe and vice versa. It is better to refer to stars as being "out." The opposite direction, "in," will be easy to establish with a plumb line which points to the center of our planet anywhere on earth. At the north pole the plumb line will indicate the north-south axis on which our earth turns counterclockwise.

High above the north pole, in our imaginary giant star bubble, lies the position of the North Celestial Pole (N.C.P.) at "plus," or "north," ninety degrees (+90°). From earth, any position in the heavens, therefore, lies south of the N.C.P. The rungs of the ladder of declination always descend southward: +90°, +80°, +70°, +60°, +50°, +40° (the declination line lying approximately above the center of the United States), +30°, +20°, +10° to zero degrees (0°), which marks the celestial equator. The ladder of declination in the heavens corresponds to the ladder of latitude on the surface of the earth, which descends from the North Pole in the arctic to the South Pole in the Antarctic. We call the lines of latitude on the earth's surface parallels. The zero position marks both the celestial equator and the earth equator beneath. Declinations in the southern celestial hemisphere are marked with minus signs, -10°, -20°, -30°, -40°, -50°, -60°, -70°, -80° and -90°, which is the position of the South Celestial Pole (S.C.P.).

Let us look at the earth "sideways" from a distant high viewpoint on the celestial equator. We see an observer on earth lying on a flatform, head to the north, feet pointing south. The observer is starframing the area straight overhead as seen from the 40th parallel in the Northern Hemisphere of the earth, for example from Philadelphia, Denver, Peking, Ankara in Turkey, Naples or Madrid, which are all close to the 40th parallel. Even though the framed area is straight overhead for the flatform observer in any of these locations, its astronomical position is 40 degrees north (+40°) (see diagram opposite). At the declination of +40 celestial degrees, the observer can see the constellation of Cygnus (page 82) in July, and Auriga (page 30) in December, as they culminate almost six months apart.

The fan-shaped list of constellations on the opposite page gives the angle of declination of the twenty-four starfields listed in this book with the dates when they culminate. Later we will show how easy it is to find these starfields in our skies. Astronomers pinpoint positions of stars, galaxies, nebulae and other heavenly bodies by giving both their east-west position in hours of right ascension (R.A.) and their north-south positions in degrees of declination (Dec.). Together these are called "coordinates."

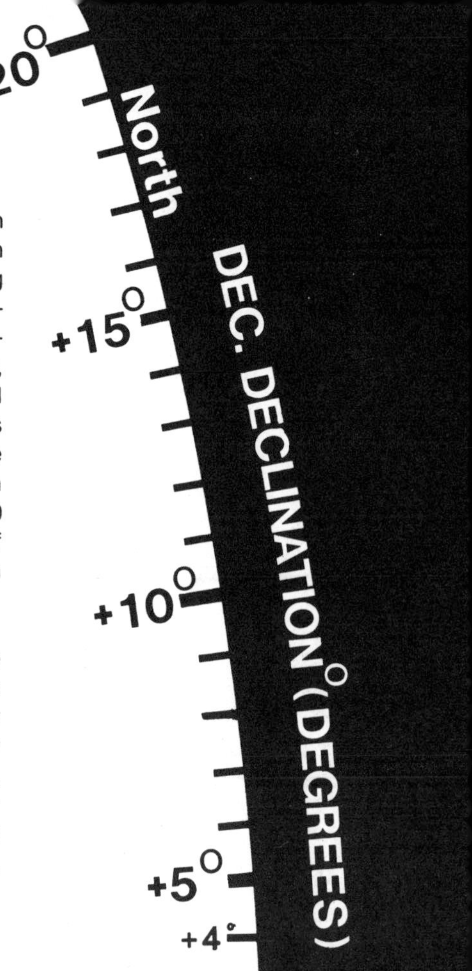

DEC. DECLINATION (DEGREES)

North

+20°
+15°
+10°
+5°
+4°
+3°
+2°
+1°

FOR ARC-MINUTES SEE PAGE 21

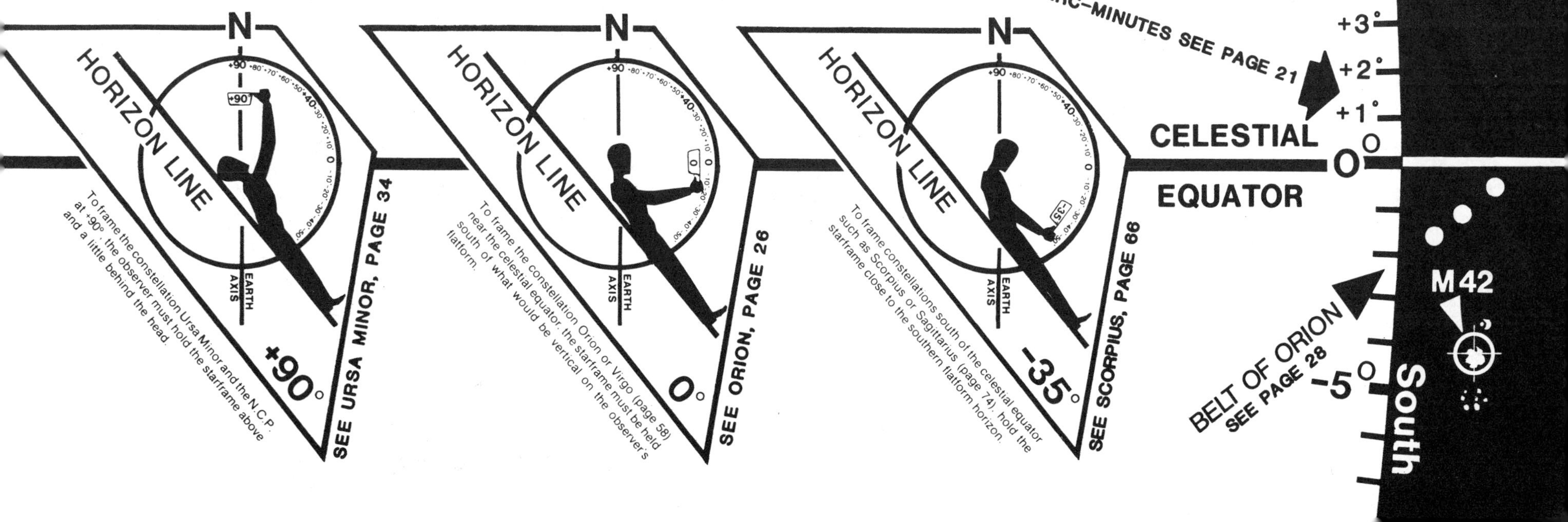

CELESTIAL
EQUATOR 0°

M 42

BELT OF ORION
SEE PAGE 28

-5°

South

HORIZON LINE

N
+90 -80 -70 -60 -50 +40 -30 -20 -10 0
+90
EARTH
AXIS

To frame the constellation Ursa Minor and the N.C.P. at +90° the observer must hold the starframe above and a little behind the head.

+90°

SEE URSA MINOR, PAGE 34

HORIZON LINE

N
+90 -80 -70 -60 -50 +40 -30 -20 -10 0
0
EARTH
AXIS

To frame the constellation Orion or Virgo (page 58) near the celestial equator, the starframe must be held south of what would be vertical on the observer's flatform.

0°

SEE ORION, PAGE 26

HORIZON LINE

N
+90 -80 -70 -60 -50 +40 -30 -20 -10 0
-35
EARTH
AXIS

To frame constellations south of the celestial equator such as Scorpius or Sagittarius (page 74), hold the starframe close to the southern flatform horizon.

-35°

SEE SCORPIUS, PAGE 66

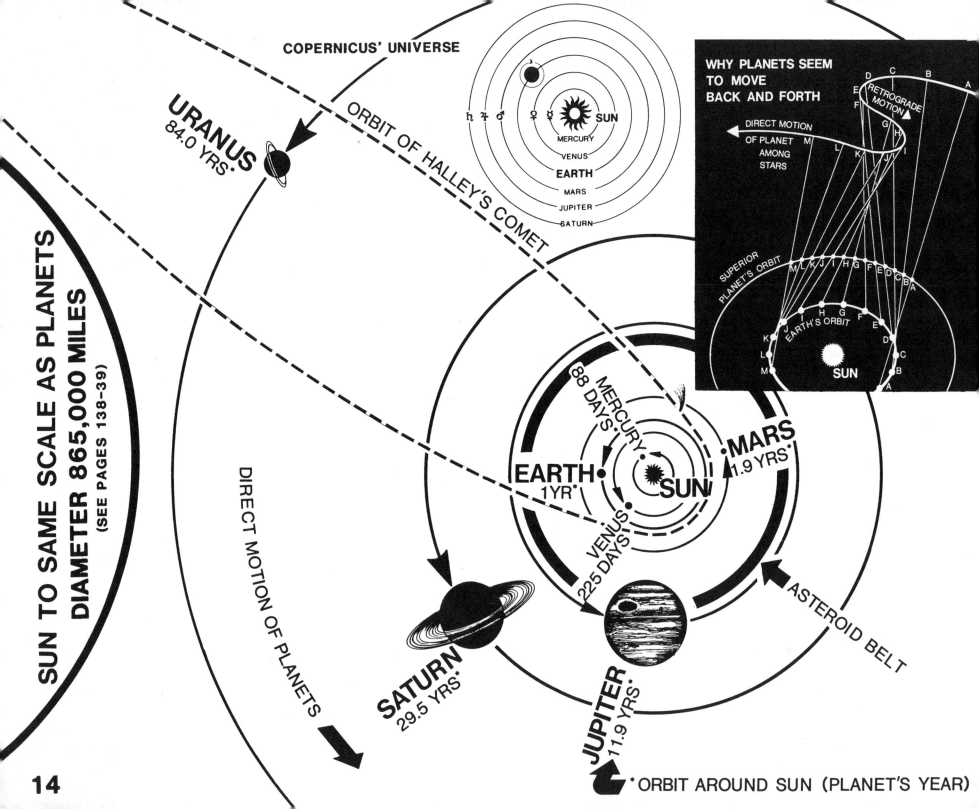

COPERNICUS' UNIVERSE

SUN
MERCURY
VENUS
EARTH
MARS
JUPITER
SATURN

WHY PLANETS SEEM TO MOVE BACK AND FORTH

RETROGRADE MOTION

DIRECT MOTION OF PLANET AMONG STARS

SUPERIOR PLANET'S ORBIT

EARTH'S ORBIT

SUN

URANUS
84.0 YRS*

ORBIT OF HALLEY'S COMET

SUN TO SAME SCALE AS PLANETS
DIAMETER 865,000 MILES
(SEE PAGES 138-39)

DIRECT MOTION OF PLANETS

MERCURY 88 DAYS*

EARTH 1 YR*

SUN

MARS 1.9 YRS*

VENUS 225 DAYS*

ASTEROID BELT

SATURN 29.5 YRS*

JUPITER 11.9 YRS*

*ORBIT AROUND SUN (PLANET'S YEAR)

14

PLUTO
247.7 YRS

PLANETS

The word means wanderers, because the positions of the planets, unlike those of stars, change constantly. The nine known solar planets, including our earth, are satellites of our sun. They are usually listed in order of increasing distance from the solar center: Mercury, Venus, Earth, Mars, Jupiter, Saturn, Uranus, Neptune and Pluto. They orbit the sun at different speeds. The time required for one revolution around the sun (one year in the case of the earth) is shown on the diagram. Planet sizes are also indicated, in relation both to the sun and to each other. In order for you to grasp the scale, this diagram would have to be expanded to fit the large sun, shown partially on the left, into the center. This would enlarge the diameter of earth's orbit to about 50 yards and the orbit of Pluto to one mile from the diagram on this page.

Mercury and Venus lie between earth and the sun; having smaller orbits than the earth's, they are called inferior planets. Because Mars, Jupiter, Saturn, Uranus, Neptune and Pluto travel in larger orbits outside our own, they are referred to as superior planets.

Throughout history, the apparent motion of the planets puzzled scholars because the usual eastward planetary journeys were interrupted by periodic westward detours. Such reversals were called "retrograde motions." Ingenious but incorrect explanations of circles within circles were proposed by the most learned men in history to try to solve the riddle of planetary motions. A careful look at the diagram opposite instantly unveils the seeming mystery of ages and clearly shows that the planets proceed in their east-west courses without interruption. It is just the view

from our earth-in-motion's vantage point which creates the "retrograde" illusion. The simple solution of such seemingly complex and puzzling problems lies at the heart of science.

The Italian Galileo Galilei, born in 1564, was the first true scientist to combine careful observations of the planet Jupiter with exceptional reasoning powers. With his newly built telescope, Galileo was able to see that Jupiter had satellites which traveled around it in predictable orbits. The famous Italian drew on earlier work of the brilliant Polish scholar Copernicus (1473-1543), who had first put the sun — not the earth — at the center of the universe. Galileo provided proof for the revolutionary idea that the earth revolved around the sun. Although his findings were correct, after he published them he was imprisoned in his home until his death, in 1642 at age seventy-eight, because his historically important conclusions defied the church.

In the diagram shown on these pages, we look down on our solar system from a distant vantage point in our galaxy. The position of the "Asteroid Belt" is indicated (see page 123). The orbit of Halley's Comet (see page 124) is shown. It is typical of the periodic comets which swing around our sun from time to time.

If we viewed this diagram sideways, along the paper on which it is printed, the orbits of the planets — but not the comet — would almost lie in a straight line parallel to the Band of the Zodiac. On close inspection we might detect a slight wobble. Like Hula-Hoops of different sizes, the orbits would move from above to below the ecliptic and back again. Mostly the planets would stay well within the Zodiac.

NEPTUNE
164.8 YRS

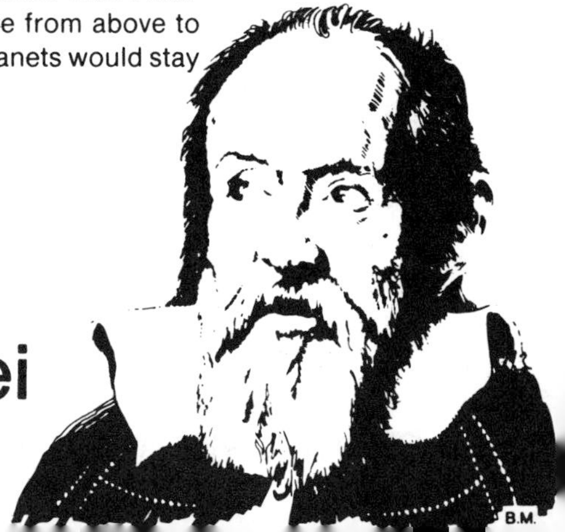

Galileo Galilei
1564 - 1642

B.M.

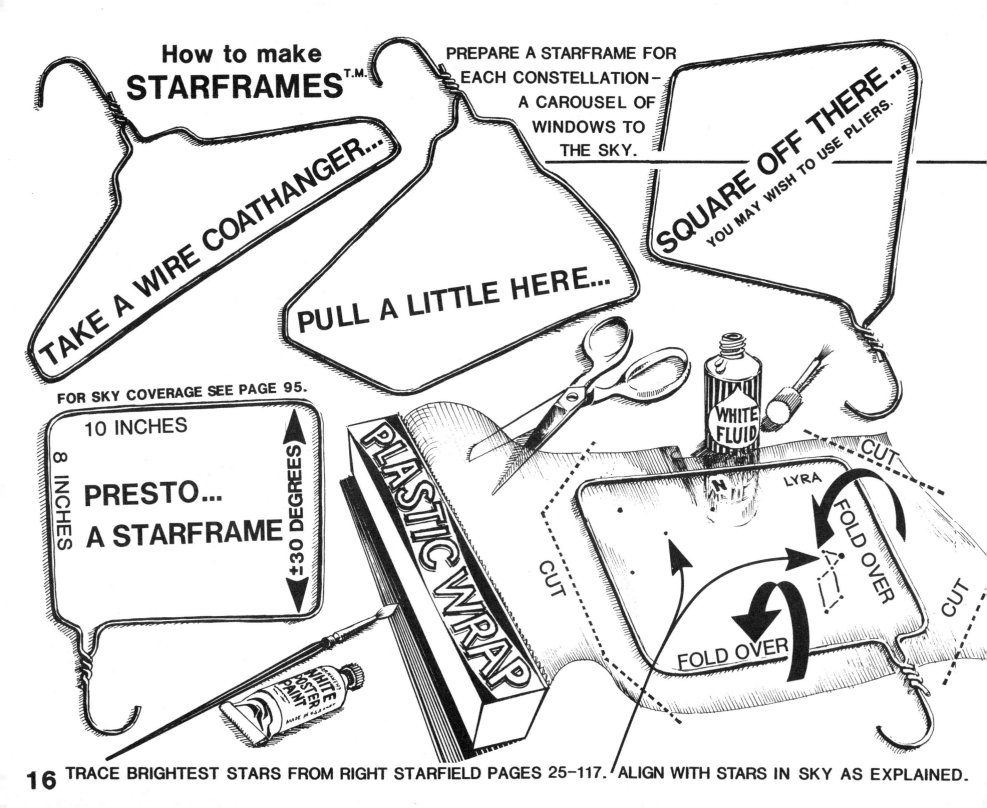

How to make
STARFRAMES™

PREPARE A STARFRAME FOR EACH CONSTELLATION — A CAROUSEL OF WINDOWS TO THE SKY.

TAKE A WIRE COATHANGER...

PULL A LITTLE HERE...

SQUARE OFF THERE...
YOU MAY WISH TO USE PLIERS.

FOR SKY COVERAGE SEE PAGE 95.

10 INCHES

8 INCHES

±30 DEGREES

PRESTO...
A STARFRAME

PLASTIC WRAP

WHITE FLUID

WHITE POSTER PAINT
MADE IN U.S.A.

CUT

CUT

CUT

FOLD OVER

FOLD OVER

LYRA

N

16 TRACE BRIGHTEST STARS FROM RIGHT STARFIELD PAGES 25–117. ALIGN WITH STARS IN SKY AS EXPLAINED.

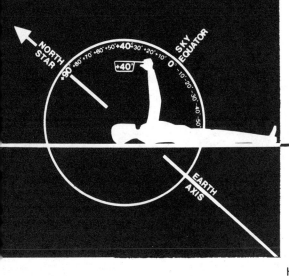

STARFRAMES

When you view constellations on or near the Band of the Zodiac, be on the lookout for very bright objects that are not shown in our starcharts. Predictably these will be planets wandering through your part of the sky. Try to remember their positions. Over a period of a few days you will be able to notice their motion in relation to the stars. While planets change from direct to retrograde motion or back again, they will appear to be almost stationary for weeks at a time (page 14).

Dress warmly; even summer nights can get chilly. Do not wait until you feel uncomfortable before putting on additional clothing. Anticipate dropping temperatures by adding layers of apparel before you feel the need for them. Avoid perspiration at all costs. Only dry fabrics can efficiently protect you from the cold. Put on extra socks. When your feet are warm, the rest of you will be comfortable, too.

Homemade starframes traced from the pages of this book present a simple new way for everyone to find celestial objects. When you know a constellation's time of culmination (times are given for twenty-four constellations in this book) all that remains for the observer to establish is the north-south direction angle of declination. When the starframe is pointed correctly and two or more of the brightest objects on the tracing are aligned with their counterparts in the sky, everything else falls into place. Once found, constellations can be memorized — even photographed — over a period of several weeks. The diagrams on the left show how to make a standard starframe. Plastic kitchen wrap or any cellophane film will serve as your transparent window to the sky when combined with the lowly wire coathanger. Make a collection, a starframe for each starfield.

With white ink, poster paint or typewriter correcting fluid, trace the principal stars and special objects of interest from the star photo that is given for each constellation in this book. Tiny dots of white can be seen in near darkness by the light of a red-covered flashlight (page 135). If you wish, you can also copy the lines connecting the stars in the constellations. You decide.

Find a comfortable place in which to lie down, always with your head to the north, feet pointing south. When in doubt about your orientation, use an inexpensive compass to help you find north. At the time of culmination hold the starframe at the angle illustrated in this book for the starfield you want to observe. The distance between the starframe and your eye will need to be established by trial and error, but for most of the starcharts in this book it will be about as shown on this page.

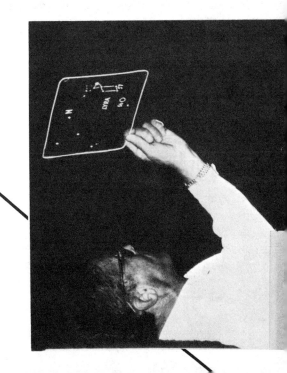

M69 M8? ?4 M47 M
04 M22 M1 M33 M8
M1 M78 M? M9 M14 M5
1 M20 M37 M?6 M19 M42
M77 M2 M34 M9 M31 M62
?6 M16 M24 M46 M54 M63
101 M73 M101 M79 M83 M
M19 M9? M84 M103 M111
148 M55 M96 M1 M81 M6
4 M28 M M19? 26 M58 M
M35 M50 ?9? ?0 M26 M

18

Hotel de Cluny, Paris.

CHARLES MESSIER

1730 - 1817

Charles Messier loved comets. Only about fifty of these were known when the French astronomer started looking for more from above the roofs of Paris, recording his observations by candlelight. Messier began his apprenticeship in 1751 at age twenty-one. His first job was as draftsman and recorder of astronomical findings. Three years later, he became a clerk at the Marine Observatory in Paris. It was housed in a residence, and observations were conducted from the octagonal tower shown below.

On January 21, 1759, Messier spotted the famous comet that Edmund Halley had predicted would return in 1758. Messier claimed to have discovered twenty-one comets of his own. As he searched the heavens for these celestial visitors, and during his other observations, he found objects which had not been seen or recorded before.

The Crab Nebula in Taurus became the first entry in the log which he kept to chart the positions of fuzzy spots in the sky, none of which he could see very clearly. His telescopes were poor even when compared to amateur instruments of today. Lenses were of low quality. Reflecting telescopes had "speculum" mirrors made of polished metal alloys, not yet of silvered glass.

All of his new discoveries must have filled Messier with great excitement and the belief that he had found new comets. Imagine his disappointment when he checked again, night after night, only to find that the blurred spots had not moved, as comets always do. In time his catalogue grew to over a hundred entries, with about twenty-five of these credited to a younger colleague named Pierre Méchain. The records cover a period from 1758 to 1790. Each finding is dated and numbered and lists positions in exact hours of right ascension and degrees of declination. There followed descriptions of the findings, many of which tell us how hard it was to resolve famed nebulae, or groupings of stars.

Since Messier's work had begun under the patronage of King Louis XVI, the approach of the bloody French Revolution affected his life in many ways. He lost his salary and his navy pension. The rent for his observatory was no longer being paid. It is said that he had to beg for the oil used in his observatory lamp. The King and aristocratic friends fell victim to the guillotine. Only with Napoleon's rise to power did Messier regain the recognition and honors he had earned. He died, childless, at the age of eighty-six.

The catalogue which Messier had prepared — partly to stop comet hunters from wasting time on hazy patches — brought immortality to his name. We place the letter "M" before the numbers which he gave to his entries. M1 is today known as the "Crab nebula." His M8 and M20 are now celebrated as the Lagoon and Trifid nebulae. M31, thought by Messier to be a nebula in Andromeda, has since been recognized as the Andromeda galaxy, an island universe just like our own Milky Way. His list of famed objects embraces some of the most beautiful sights in the night sky.

Just like Messier himself, you may only be able to see some of these treasures very vaguely for a start. In time you may want to explore them further.

Medieval starcharts engraved in copper

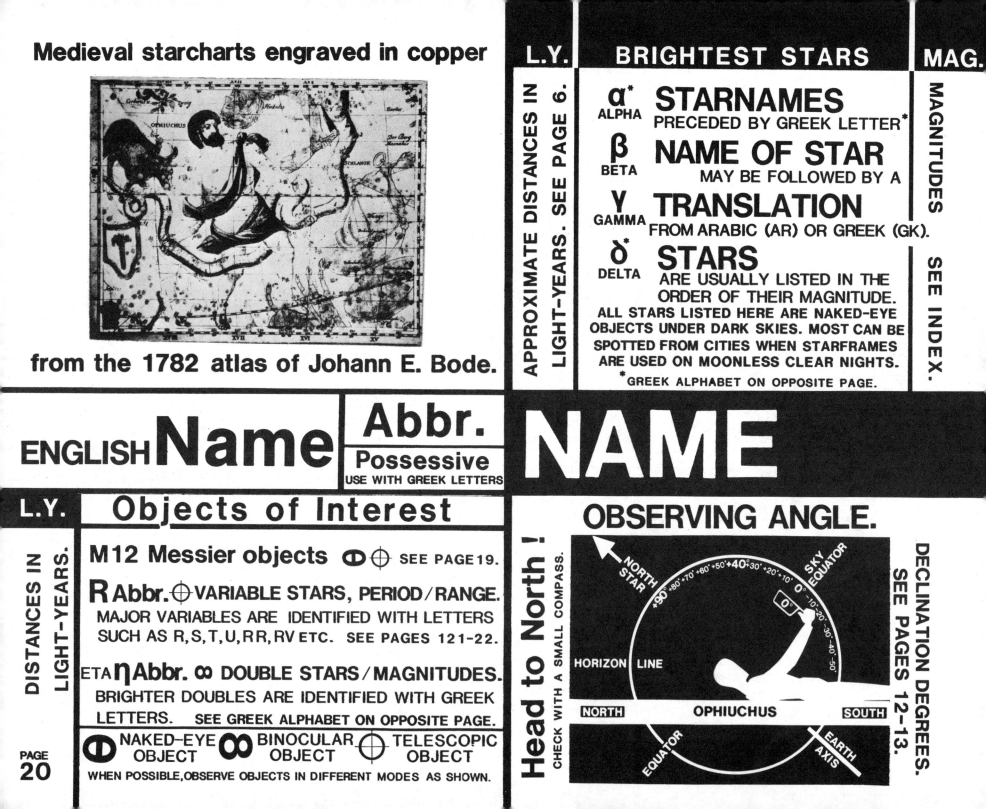

from the 1782 atlas of Johann E. Bode.

L.Y. BRIGHTEST STARS MAG.

APPROXIMATE DISTANCES IN LIGHT-YEARS. SEE PAGE 6.

MAGNITUDES SEE INDEX.

α* ALPHA — **STARNAMES** PRECEDED BY GREEK LETTER*

β BETA — **NAME OF STAR** MAY BE FOLLOWED BY A

γ GAMMA — **TRANSLATION** FROM ARABIC (AR) OR GREEK (GK).

δ DELTA — **STARS** ARE USUALLY LISTED IN THE ORDER OF THEIR MAGNITUDE.

ALL STARS LISTED HERE ARE NAKED-EYE OBJECTS UNDER DARK SKIES. MOST CAN BE SPOTTED FROM CITIES WHEN STARFRAMES ARE USED ON MOONLESS CLEAR NIGHTS.

*GREEK ALPHABET ON OPPOSITE PAGE.

ENGLISH **Name** | **Abbr.** Possessive USE WITH GREEK LETTERS

NAME

L.Y. Objects of Interest

DISTANCES IN LIGHT-YEARS.

M12 Messier objects ⊕ ⊕ SEE PAGE 19.

R Abbr. ⊕ VARIABLE STARS, PERIOD / RANGE. MAJOR VARIABLES ARE IDENTIFIED WITH LETTERS SUCH AS R, S, T, U, RR, RV ETC. SEE PAGES 121-22.

ETA η Abbr. ∞ DOUBLE STARS / MAGNITUDES. BRIGHTER DOUBLES ARE IDENTIFIED WITH GREEK LETTERS. SEE GREEK ALPHABET ON OPPOSITE PAGE.

⊕ NAKED-EYE OBJECT ∞ BINOCULAR OBJECT ⊕ TELESCOPIC OBJECT

WHEN POSSIBLE, OBSERVE OBJECTS IN DIFFERENT MODES AS SHOWN.

OBSERVING ANGLE.

Head to North !

CHECK WITH A SMALL COMPASS.

NORTH STAR

NORTH

+90° +80° +70° +60° +50° +40° +30° +20° +10° 0°

SKY EQUATOR

0° −10° −20° −30° −40° −50°

0°

HORIZON LINE

NORTH — OPHIUCHUS — SOUTH

EQUATOR

EARTH AXIS

DECLINATION DEGREES. SEE PAGES 12-13.

19HR. 17HR. 16HR. 15HR.
20HR. 18HR. 14HR. 13HR.
LYR SGR HER SCO LIB VIR
CYG SOUTH

EARTH IN MOTION
SEE PAGES 10–11

NORTH

MIDNIGHT CULMINATION

GUIDE TO STARCHARTS

1 JUN. MIDNIGHT
15 JUN. 11 P.M.
1 JUL. 10 P.M. ◄ OBSERVING DATES/TIMES
15 JUL. 9 P.M.
1 AUG. 8 P.M.

FOR DAYLIGHT SAVING TIME, SEE NOTE ON PAGE 10.

SEE OPPOSITE PAGE FOR OBSERVING ANGLE.

This pair of pages explains the format used for the starcharts in this book. Historical reproductions from the star atlas published by Johann Elert Bode in 1782 show the traditional view of each constellation. Bode was a noted German astronomer whose atlas covered the entire sky. We have reproduced Bode's renderings in the top left corner. Where coordinates can be seen on Bode's charts they correspond to the ones which are given for the brightest star of each constellation on the following starcharts. For the guidance of astrophotographers, we have marked the general target points for each pair of the twenty-four starphotos. The exact position is pinpointed where an ordinary 35mm single-lens reflex camera (SLR with standard or wide-angle lens as noted) was aimed to collect our starlight. Any amateur can match these basic photographs (pages 127-29).

The name for each constellation is on the left page, flanked, where applicable, by the appropriate ancient symbols on the right and the abbreviation and possessive form of the word on the left. The English translation of the name completes the crossbar.

The box in the top right opposite lists a maximum of eight brightest stars in each area, most of which should be visible even from light-polluted cities. Their approximate distances from us are given in light-years, also their magnitudes rounded to the tenth. The meaning of starnames is spelled out where known, translated from the Arabic (AR) or from the Greek (GK).

The Greek letters which astronomers use to identify stars (almost always in order of their brightness) are shown and named. By using the appropriate Greek letter in combination with the possessive form of the name of the constellation, you will im-

mediately be able to correctly identify principal stars. Thus, you will be able to recognize the brightest star in Taurus (page 22), named "Aldebaran," as Alpha (α) Tauri, its proper scientific name.

The box in the bottom left gives a maximum of eight objects of interest in each constellation. These may appear starlike, and only a few are visible to the naked eye. Most require binoculars or a telescope. Many of the special objects will readily reveal their position, shape and color when photographed with simple cameras (see pages 127-31). *All* the Greek-lettered stars listed will record in short-exposure photographs. A simple telescope is ideal for viewing clusters, nebulae or double stars. Some of the special objects taken through a large telescope with long time exposures are shown below and on the cover. These are to the same scale and have "north up," which is how they would appear through binoculars. Viewed through telescopes which invert images, they will appear upside down. For such observations, rotate the book so that the images will match what you see in the eyepiece.

Since the constellations in this book are presented in the order in which they culminate, giving dates and times when they are best observed at their highest in the sky, only their declination (north-south position) needs to be illustrated. Diagrams clearly explain the angle at which the starframe (or the camera) must be aimed by an observer whose head is to the north, feet pointing south. For a flatform, you may wish to use the roof of some house, or, better yet, a meadow in a dark area. Take a little compass, because you always need to know where north is, and blankets to keep warm.

THE GREEK ALPHABET

α	ALPHA	η	ETA	ν	NU	τ	TAU
β	BETA	θ	THETA	ξ	XI	υ	UPSILON
γ	GAMMA	ι	IOTA	ο	OMICRON	φ	PHI
δ	DELTA	κ	KAPPA	π	PI	χ	CHI
ε	EPSILON	λ	LAMBDA	ρ	RHO	ψ	PSI
ζ	ZETA	μ	MU	σ	SIGMA	ω	OMEGA

HIGH CONTRAST TELESCOPE PHOTOGRAPHS OF PRINCIPAL OBJECTS TO SCALE SHOWN. COLOR-PRINTS OF THESE 30-90 MINUTE EXPOSURES ARE PRESENTED ON THE COVER OF THIS BOOK. SUCH PHOTOGRAPHY REQUIRES PERFECT POLAR ALIGNMENT WITH MOTORIZED "GUIDED" SYSTEMS.

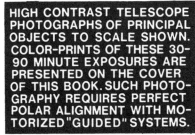

25 ARC-MINUTES

10'
+2°
50'
40'
30'
20'
10'
+1°
50'
40'
30'
20'
10'
0°

1 DEGREE = 60 ARC-MINUTES.
(1 ARC-MINUTE = 60 ARC-SECONDS)

WHAT'S UP? SEE PAGE 13

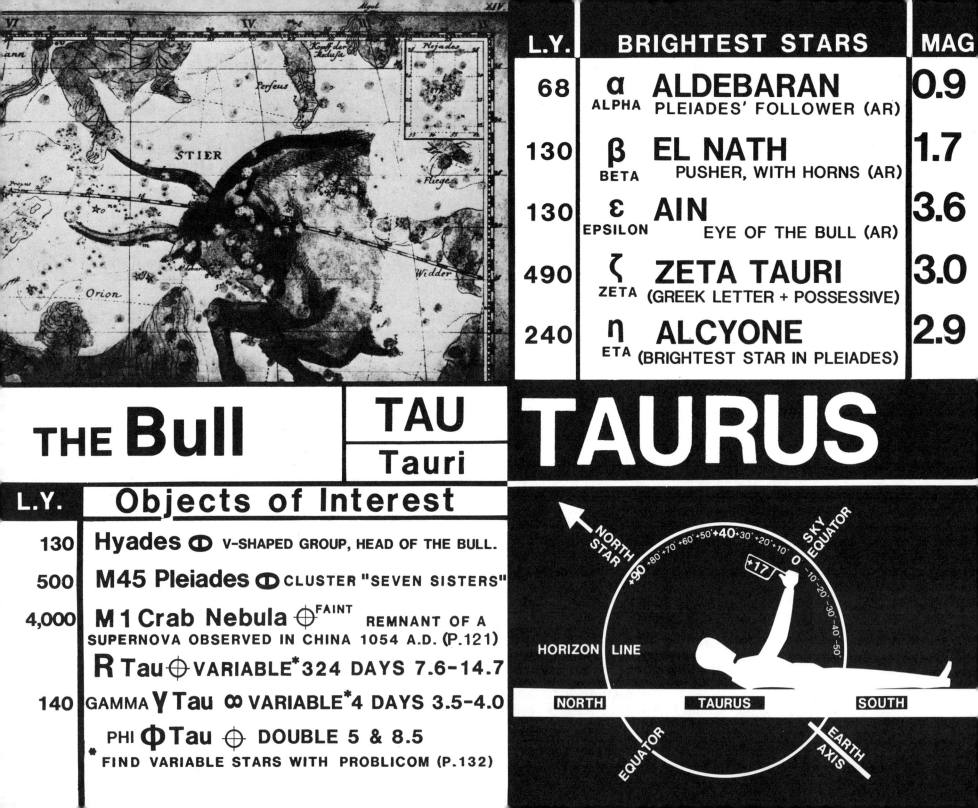

L.Y.	BRIGHTEST STARS		MAG
68	α ALPHA	ALDEBARAN PLEIADES' FOLLOWER (AR)	0.9
130	β BETA	EL NATH PUSHER, WITH HORNS (AR)	1.7
130	ε EPSILON	AIN EYE OF THE BULL (AR)	3.6
490	ζ ZETA	ZETA TAURI (GREEK LETTER + POSSESSIVE)	3.0
240	η ETA	ALCYONE (BRIGHTEST STAR IN PLEIADES)	2.9

THE Bull

TAU
Tauri

TAURUS

Objects of Interest

L.Y.		
130	Hyades	V-SHAPED GROUP, HEAD OF THE BULL.
500	M45 Pleiades	CLUSTER "SEVEN SISTERS"
4,000	M1 Crab Nebula	FAINT REMNANT OF A SUPERNOVA OBSERVED IN CHINA 1054 A.D. (P.121)
	R Tau	VARIABLE* 324 DAYS 7.6–14.7
140	GAMMA γ Tau	∞ VARIABLE* 4 DAYS 3.5–4.0
	PHI Φ Tau	DOUBLE 5 & 8.5

* FIND VARIABLE STARS WITH PROBLICOM (P.132)

NORTH STAR

+90 +80 +70 +60 +50 +40 +30 +20 +10 0 SKY EQUATOR

-10 -20 -30 -40 -50

+17

HORIZON LINE

NORTH   TAURUS   SOUTH

EQUATOR

EARTH AXIS

1 DEC. MIDNIGHT
15 DEC. 11 P.M.
1 JAN. 10 P.M.
15 JAN. 9 P.M.
1 FEB. 8 P.M.

Visit your library and check out the astronomy section. The nights may be cold now, but Newton's laws, NASA flights, telescope building and total eclipses all await you in book or audiovisual form. Check out a star atlas and compare its pages with the photographs shown in this book. You will be able to make armchair discoveries of your own.

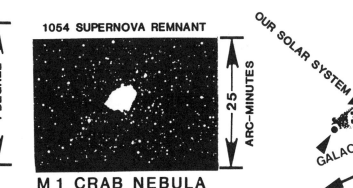

PATH OF THE SUN.

TAU

M 45

ALDEBARAN

TAURUS

Taurus, the Bull, is one of the twelve constellations of the Zodiac. The sun travels through it from May 14 to June 21. Observations at sunrise and sunset allowed early astronomers to determine that even though the stars and the constellations are not visible to us in daytime, they were still out there on the celestial sphere. The apparent path of the sun (the ecliptic) and the Band of the Zodiac girdle our planet day and night. Midnight is the time when the sun is at its highest point on the opposite side of our planet.

The sparkling Pleiades are a famous part of Taurus and should help locate our starfield quite easily at the times listed above. In this grouping, sometimes called the Seven Sisters, the average naked eye can see about six or seven stars. With a simple optical aid such as binoculars or a basic telescope, ten times as many may be spotted. M45 is the number which Messier assigned to this most famous "galactic cluster." Such loose, or "open" groupings all lie within the Frisbee shaped disc of our Milky Way galaxy near what is called our "galactic plane."

The lines superimposed on stars to mark the constellations are not always easy to recognize or to relate to a theme. The bull is one of the clearer pictograms, with four distinguishable legs and over-sized horns. The V-shaped head of Taurus is usually referred to as the Hyades, a relatively nearby cluster. Like the Pleiades, they relate to the fanciful Greek myths that surround many heavenly objects. Since these pages are concerned with getting to know the sky, we will not dwell on mythology. It may be enough to say that the powerful and restless bull was an important part of ancient cultures, both in religion and in magic.

A more practical observation of the Hyades has led to their being called the rainy stars. Their rising in the east in autumn introduced the rainy season in Mediterranean climates.

Aldebaran is the brightest star in Taurus. At +17° Dec. it is well situated for convenient observation from our latitude. You can also refer to it as Alpha (α) Tauri (Greek letter plus possessive form). Beta (β) Tauri is off the chart at the top left. This star, marking the tip of the northern horn, is shared by Taurus and Auriga (page 32), where it forms the southernmost point of the charioteer's pentagon.

The Crab Nebula, M1, shown below for scale, not brightness, is a very faint object and needs dark skies and a telescope. It is all that remains of the explosion of a star, a supernova, (page 120) that occurred over nine hundred years ago. Chinese records tell of a "guest star" which was so bright it could be seen in daytime. So cataclysmic was this stellar explosion that even today the remaining cloud of gas expands at the rate of some 1,000 miles per second.

M45 PLEIADES GALACTIC CLUSTER

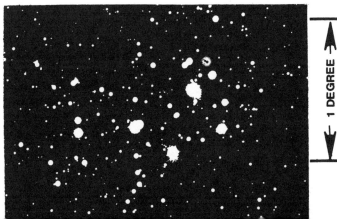

1 DEGREE

1054 SUPERNOVA REMNANT

25 ARC-MINUTES

M 1 CRAB NEBULA

OUR GALAXY (THE MILKY WAY)

OUR SOLAR SYSTEM

GALACTIC (OPEN) CLUSTERS

100,000 LIGHTYEARS DIAMETER

23

Find the pictogram stars on this page and connect them with ruled colored lines. You can also create your own ideograms with red markings which will filter out and disappear under a red flashlight (page 135).

GEMINI PAGE 40

4h

ARIES PAGE 112

BINOCULAR FIELD OF VIEW

ALCYONE

M 45 PLEIADES

φ

τ υ

κ

η

ζ

M 1

PATH OF THE SUN IN MAY / JUNE

ω

ECLIPTIC

ε

δ

HYADES

ALDEBARAN α

σ θ γ

TAURUS

α Tau 04h36m +16°31'

NOTE: THIS IS A "NEGATIVE" PRINT MADE FROM A COLOR SLIDE. IMAGES ON THE RIGHT PAGES SHOW THE STARLIGHT WHICH CAMERAS CAN COLLECT IN SECONDS — IN COLOR — FOR SLIDE PROJECTION. THIS WILL TURN ANY ROOM INTO A PLANETARIUM.

λ

+10°

ξ

μ

R

ο

ORION PAGE 28

ν

LONG EXPOSURE PHOTOGRAPHS REACHING FAINT STARS

ALIGN WITH ALDEBARAN AND PLEIADES

(HOLD STARFRAME TO FIND AND OVERLAP
THE BRIGHTEST OBJECTS IN THIS STARFIELD)

N

STANDARD 50 MM LENS

PLEIADES

HYADES

ALDEBARAN
(ALPHA)

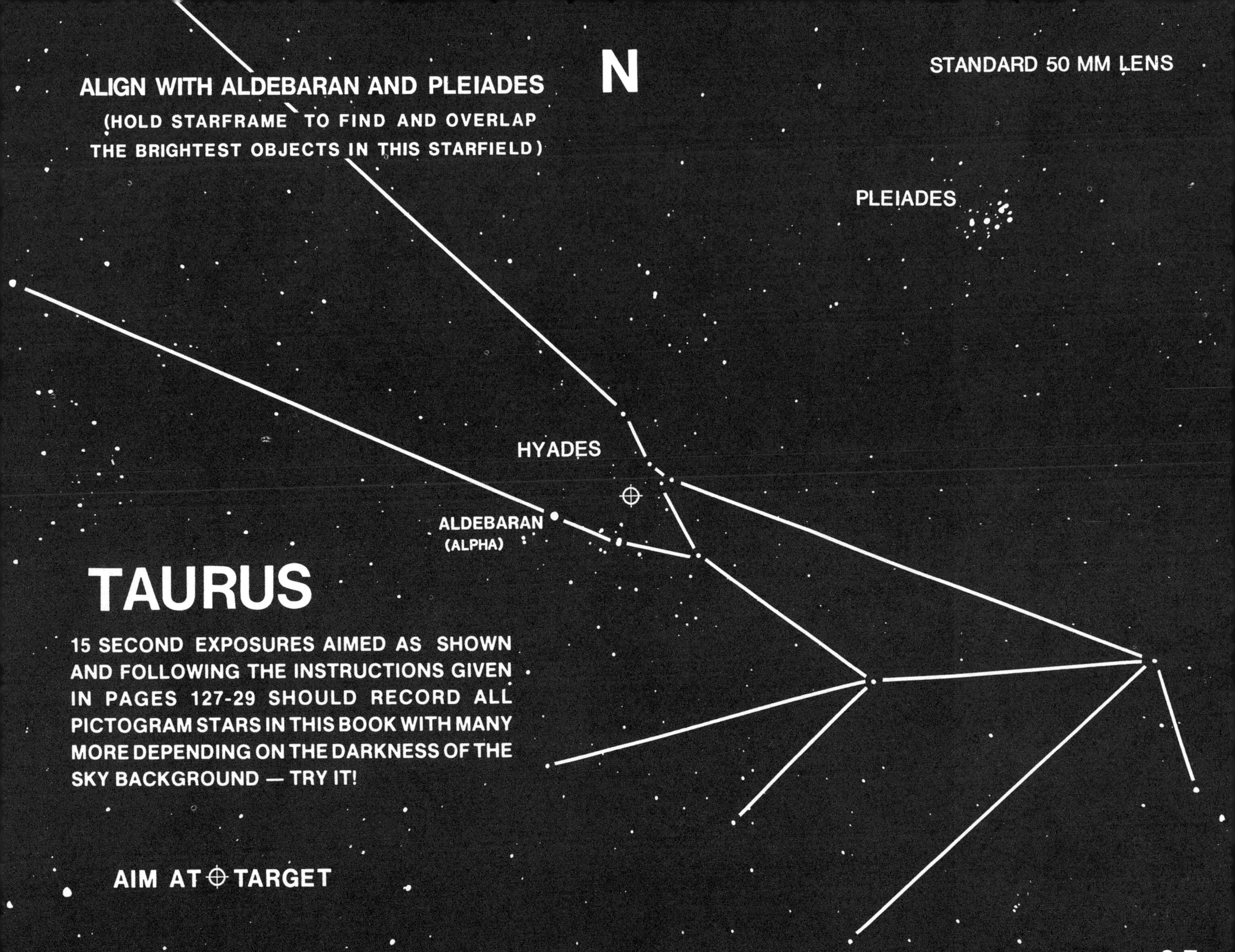

TAURUS

15 SECOND EXPOSURES AIMED AS SHOWN
AND FOLLOWING THE INSTRUCTIONS GIVEN
IN PAGES 127-29 SHOULD RECORD ALL
PICTOGRAM STARS IN THIS BOOK WITH MANY
MORE DEPENDING ON THE DARKNESS OF THE
SKY BACKGROUND — TRY IT!

AIM AT ⊕ TARGET

SAME AREAS IN SHORT EXPOSURE PHOTOGRAPHS SHOW FEWER STARS

THE Hunter

ORI
Orionis

ORION

Objects of Interest

L.Y.		
2,000	M42-43 ∞⊕	Orion nebula
1,600	M78 ⊕ FAINT	Reflection nebula
1,200	IC434 ⊕ VERY FAINT	Horsehead nebula
	S Ori VARIABLE	416 DAYS 7.5-13.5
910	BETA β Ori ⊕	DOUBLE 0.1 & 7.0
1,800	LAMBDA λ Ori ⊕	DOUBLE 3.7 & 5.6

NORTH STAR

+90 +80 +70 +60 +50 +40 +30 +20 +10 0 SKY EQUATOR -10 -20 -30 -40 -50

+02

HORIZON LINE

| NORTH | ORION | SOUTH |

EQUATOR

EARTH AXIS

15 DEC. MIDNIGHT
1 JAN. 11 P.M.
15 JAN. 10 P.M.
1 FEB. 9 P.M.
15 FEB. 8 P.M.

If you live near a planetarium, why not go and see a star show on a cold wet night in warm comfort? Your starframes will help you find your way even under projected night skies. Phone ahead, find out which constellations will be shown. Then prepare to look for your favorite stars.

Orion is easy to find and impossible to forget once you have seen it. Known as the Hunter, the Warrior, or even as the Giant, it lies partly in the winter Milky Way. Even if you cannot see the faintest stars of our galaxy, the familiar belt of the legendary swordsman will be visible from cities. Just look for its distinctive three bright stars.

The constellation straddles the celestial equator and can therefore be seen from all populated areas of the Northern and Southern Hemispheres.

Orion contains spectacular treasures. None can compare with the great Orion nebula, given the numbers 42 and 43 by Charles Messier. This gigantic cloud of glowing gas can be detected with the naked eye as a hazy object. Binoculars reveal it as a swirling nebula near the head of the "dagger" of Orion. A small telescope will resolve what seems in binoculars like one star near the center of M42 into four close but separate components. These are called the "Trapezium."

Orion's caldron of gas is a birthplace of new stars. At this very moment masses of stellar dust are being drawn together here by the merest traces of gravity and are beginning to fall in on themselves. They are starting the genesis process of star formation. In time they will kindle vast balls of compressed nuclear fires, origins of light-giving suns.

To the upper left of the Orion nebula lies M78. Unlike M42, which glows from within, this nebula only reflects the light of a nearby star. The distinctive Horsehead nebula lies just to the south. Its towering dark clouds of stardust are not easily seen, even with large telescopes, but cameras with light-sensitive films can collect the starglow behind this faint silhouette.

The star Betelgeuse, Alpha (α) Orionis, has a distinctly orange color and belongs to the so-called red giants. It is the eleventh brightest star in the heavens.

Rigel, Beta (β) Orionis, though identified with only the second letter of the Greek alphabet, is really the brightest star in Orion. It blazes with a bluish-white brilliance and is the seventh brightest among all the stars in our sky. With a medium telescope of 5-inch diameter, Rigel may be seen to be a double star. There can be no turbulence in the air, and "seeing" must be at its very best. Otherwise the companion, at faint magnitude 7, will be lost in the glare of the near-zero magnitude of the supergiant.

Mintaka, Delta (δ) Orionis, at magnitude 2.2, also has a companion of 7th magnitude lying just north of its parent star. It may be easier to see because a greater distance separates the pair.

With fast color films, a photograph of the constellation Orion overall makes a wonderful picture. The glowing nebulae will quickly reveal themselves in rosy pinks. That is the color signature of the condensed hydrogen gas, the most common of the elements in the universe.

HORSEHEAD NEBULA

IC 434

BIRTHPLACE OF NEW STARS

M 42   ORION NEBULA

REFLECTION NEBULA

M 78

Find the pictogram stars on this page and connect
them with ruled colored lines. You can also create your
own ideograms with red markings which will filter out
and disappear under a red flashlight (page 135).

GEMINI PAGE 40

AURIGA PAGE 32

TAURUS PAGE 24

μ

{λ

BETELGEUSE ● α

γ

M78

O° ⊙   SKY (CELESTIAL) EQUATOR   δ

ε

IC 434 ► ⊕ ζ

η

ORION

α Ori 05h55m +7°24'

CANIS MAJOR PAGE 44

M42/43   ► S

RIGEL ● β

K

28   "BELT" OF ORION LIES CLOSE TO CELESTIAL EQUATOR

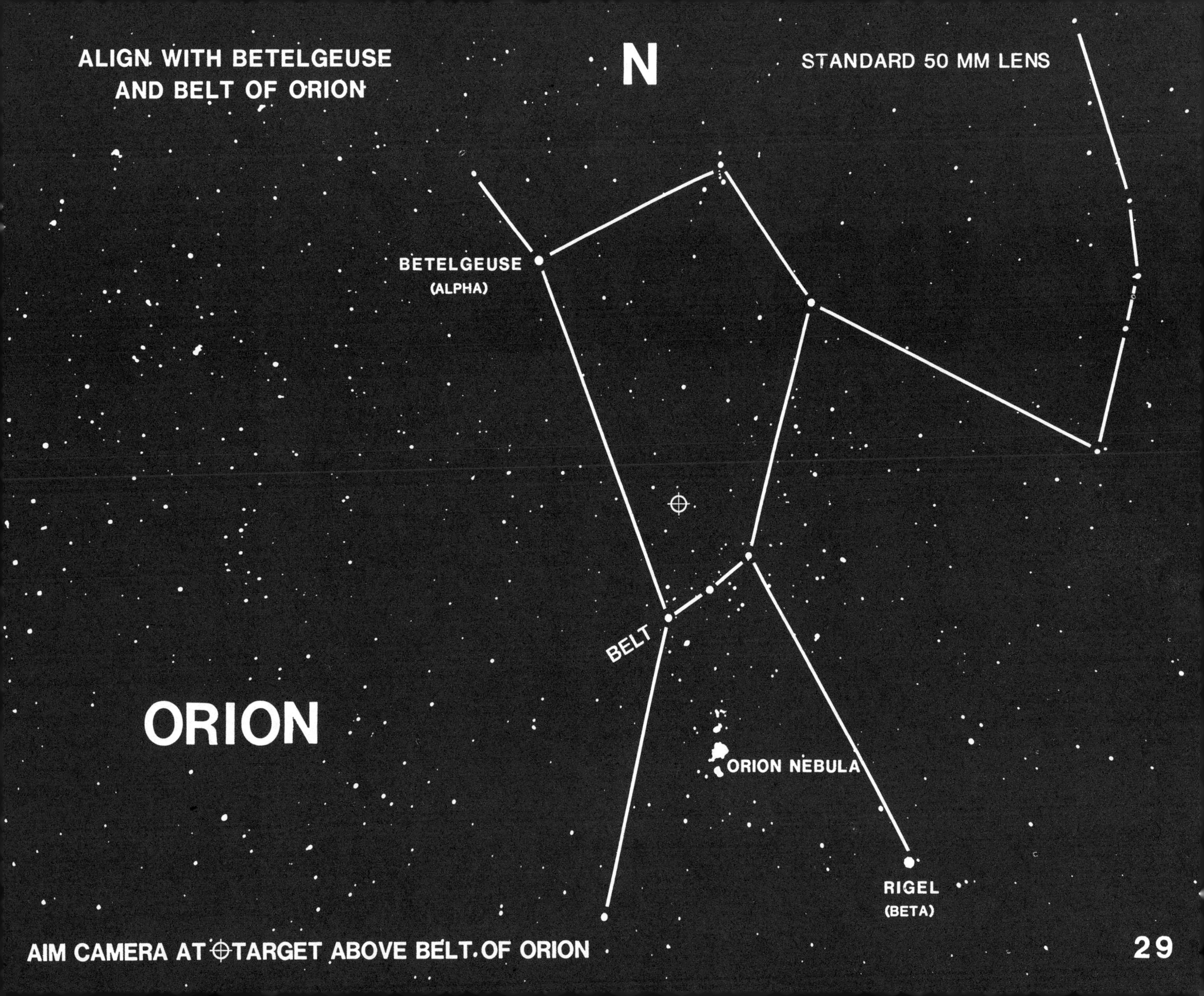

ALIGN WITH BETELGEUSE
AND BELT OF ORION

N

STANDARD 50 MM LENS

BETELGEUSE
(ALPHA)

ORION

BELT

ORION NEBULA

RIGEL
(BETA)

AIM CAMERA AT ⊕ TARGET ABOVE BELT OF ORION

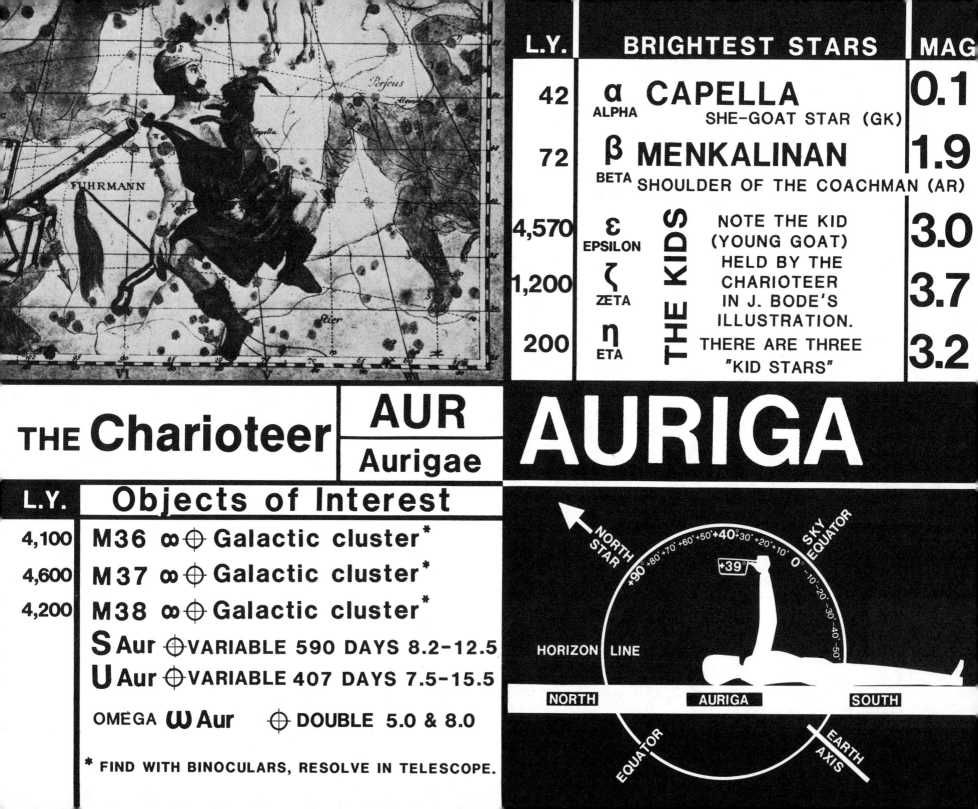

L.Y.	BRIGHTEST STARS		MAG
42	α **CAPELLA** ALPHA	SHE-GOAT STAR (GK)	**0.1**
72	β **MENKALINAN** BETA	SHOULDER OF THE COACHMAN (AR)	**1.9**
4,570	ε EPSILON	NOTE THE KID (YOUNG GOAT) HELD BY THE CHARIOTEER IN J. BODE'S ILLUSTRATION. THERE ARE THREE "KID STARS"	**3.0**
1,200	ζ ZETA		**3.7**
200	η ETA		**3.2**

THE KIDS

THE Charioteer

AUR

Aurigae

AURIGA

L.Y.	Objects of Interest
4,100	**M36** ∞ ⊕ **Galactic cluster***
4,600	**M37** ∞ ⊕ **Galactic cluster***
4,200	**M38** ∞ ⊕ **Galactic cluster***
	S Aur ⊕ VARIABLE 590 DAYS 8.2-12.5
	U Aur ⊕ VARIABLE 407 DAYS 7.5-15.5
	OMEGA ω Aur ⊕ DOUBLE 5.0 & 8.0

* FIND WITH BINOCULARS, RESOLVE IN TELESCOPE.

NORTH STAR

+90° +80° +70° +60° +50° **+40°** +30° +20° +10° 0° -10° -20° -30° -40° -50°

SKY EQUATOR

+39°

HORIZON LINE

| NORTH | AURIGA | SOUTH |

EQUATOR

EARTH AXIS

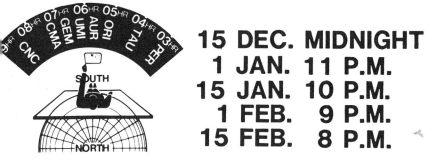

15 DEC. MIDNIGHT
1 JAN. 11 P.M.
15 JAN. 10 P.M.
1 FEB. 9 P.M.
15 FEB. 8 P.M.

Now is a good time to build a Problicom (pages 132-33) with a friend who may own the same-model projector as you do. Friends who share a Problicom and "blink" together make more discoveries. Invite a group when you project your photographs, and let them share the beauty and color which your star pictures will readily show.

AURIGA

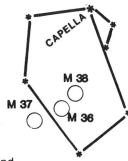

The constellation Auriga lies in the same right-ascension (R.A.) hour slice of the celestial sphere as does Orion. The declinations of the two are different. Orion, with its distinctive "belt" at 0°, lies near the celestial equator, while Auriga's pentagon, at +40° Dec., passes nearly overhead for 40th parallel observers in the Northern Hemisphere.

Capella, the brightest star in the constellation of the "Charioteer," is a double star. It is a so-called "spectroscopic binary," which means that the pair can be detected only by analyzing their starlight with special instruments. The two stars are 70 million miles apart and revolve around a common center of gravity.

Capella is the northernmost star of the celestial pentagon, which is said to make up the body of a wagon driver. Don't waste time searching for the image of a coachman. It will prove equally useless to try to find the "she-goat" which Alpha (α) Aur is supposed to represent. The copperplate image in Bode's sky atlas will have to serve as a substitute for a legible pictogram. The southernmost point in the five-cornered shape is El Nath, Beta (β) Tau, a star shared by Auriga and Taurus.

Two objects listed in the catalog of Charles Messier can be found within the pentagon. Both M36 and M38 appear bright as 6th-magnitude stars, which puts them at the limit of what the naked eye can see. Binoculars or an inexpensive telescope will reveal galactic clusters. There is a third such open grouping, M37, a little farther south and to the left. Each of these three Messier objects consists of about a hundred stars.

T Aurigae is a nova discovered in the constellation in the year 1891. A nova is a star which suddenly undergoes explosive changes, when it becomes hundreds of thousands of times brighter. After throwing off its outer layers, the star goes back to its fainter state. This makes novae "variable stars" (page 121). T Aur blazed forth by brightening from dim magnitude 15 to where it could be seen with the naked eye at magnitude 4. Like all such irregular variables, this star could blast off again at almost any time. You may want to photograph and "blink" the area (pages 132-33) to keep an eye on it.

GALACTIC CLUSTER

M 36

GALACTIC CLUSTER

M 37

GALACTIC CLUSTER

M 38

CAPELLA "B"

ONE REVOLUTION IN 104 DAYS

70 MILLION MILES

CAPELLA "A"

►Size of our SUN for scale

31

AURIGA
α Aur 05h 17m +45°59'

Find the pictogram stars on this page and connect them with ruled colored lines. You can also create your own ideograms with red markings which will filter out and disappear under a red flashlight (page 135).

PERSEUS PAGE 116

MENKALINAN • β

CAPELLA • α

ε

+40°

η ζ

γ

ν • τ

θ • •

u

ω

M38

M36

M37

S

I

GEMINI PAGE 40

ORION PAGE 28

TAURUS PAGE 24

U T

β Tau = γ Aur

NOVA 1891

32

AURIGA

MENKALINAN
(BETA)

CAPELLA
(ALPHA)

L.Y.	BRIGHTEST STARS		MAG
150	α ALPHA	**POLARIS** NORTH STAR (LATIN) ALSO KNOWN AS STELLA POLARIS	2.0
95	β BETA	**KOCHAB** STAR (HEBREW)	2.0
230	γ GAMMA	**PHERKAD** TWO CALVES (WITH β) (AR)	3.0
230	δ DELTA	**YILDUN** (DELTA URSAE MINORIS)	4.4
300	ε	**EPSILON UMI**	4.0

LITTLE Bear

UMI
Ursae Minoris

URSA MINOR

Objects of Interest

L.Y.

THERE ARE NO MESSIER OBJECTS IN UMI.

α **Alpha Ursae Minoris (Polaris)** ⊕ IS BOTH A VARIABLE STAR AND A DOUBLE. CHANGES IN MAGNITUDE CAN BE PERCEIVED BY TRAINED A.A.V.S.O. MEMBERS. SEE P. 121. PERIOD: 3.96 DAYS. RANGE: 1.92-2.07

THE COMPANION STAR IS OF 9TH MAG.

S UMi ⊕ VARIABLE 326 DAYS 7.0-12.4

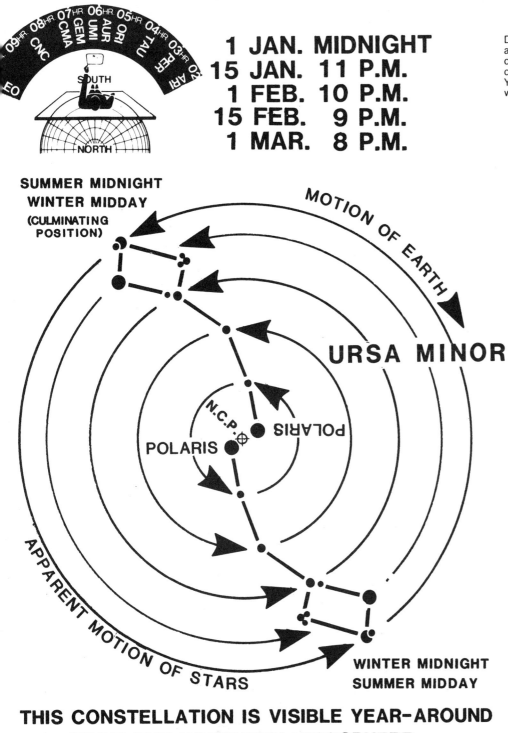

1 JAN. MIDNIGHT
15 JAN. 11 P.M.
1 FEB. 10 P.M.
15 FEB. 9 P.M.
1 MAR. 8 P.M.

SUMMER MIDNIGHT
WINTER MIDDAY
(CULMINATING POSITION)

MOTION OF EARTH

URSA MINOR

N.C.P.

POLARIS

POLARIS

APPARENT MOTION OF STARS

WINTER MIDNIGHT
SUMMER MIDDAY

THIS CONSTELLATION IS VISIBLE YEAR-AROUND FROM THE NORTHERN HEMISPHERE

Daily weather-satellite photographs in newspapers allow anyone to become familiar with air currents and cloud cover. Collect these invaluable records and study them in connection with the forecast maps usually printed nearby. You too can become a weatherperson to tell whether you will be able to see the stars tonight.

This is as far north as you will have to strain your back to look for a constellation. Because it lies so close to the North Celestial Pole, Ursa Minor and other constellations near the N.C.P. are called "circumpolar," meaning "around the pole," and are visible to us throughout the year.

The Little Bear has Polaris, the North Star, as its brightest star. Its scientific name is Alpha (α) Ursae Minoris (Greek letter plus possessive form). It lies less than one degree from the N.C.P. (page 36). Our starchart shows the constellation in its winter position. The diagram here presents both the winter and the summer positions. Each shows the Little Bear, also known as the Little Dipper, just south of the N.C.P.

In the summer midnight position, the dipper is upside down and is empty after "going over the top." Ursa Minor is the only starfield we will show sneaking "under" the North Star. It would be impossible to do so with Orion or Auriga. Both would soon set behind the horizon and be hidden from view. You may want to review pages 10 and 11, which can help explain why tomorrow at noon Ursa Minor will be in a position where its dipper will be in the right-side-up position, with the handle near the top. We will not be able to see the stars, but they will be out there just the same. It will help to remember that we on earth will have made half a daily turn on our polar axis by then. In relation to the celestial twenty-four-hour clock of right ascension, the Little Dipper will not have moved. It will stay fixed in its fifteen-hours R.A. position even when seen half a yearly revolution hence, six months from now.

Ursa Minor is a wonderful year-round celestial-constellation clock with which to study the motion of our planet. Watch or photograph it during the next twelve months, and get a good grasp on our celestial sphere as it relates to earth.

Polaris is a double star. The companion of 9th magnitude can be seen with telescopes. Although the two stars seem very close together, the distance separating the pair is at least two thousand times as great as the one which separates the earth from the sun. Beta (β) UMi and Gamma (γ) UMi are sometimes referred to as the "guardians of the pole."

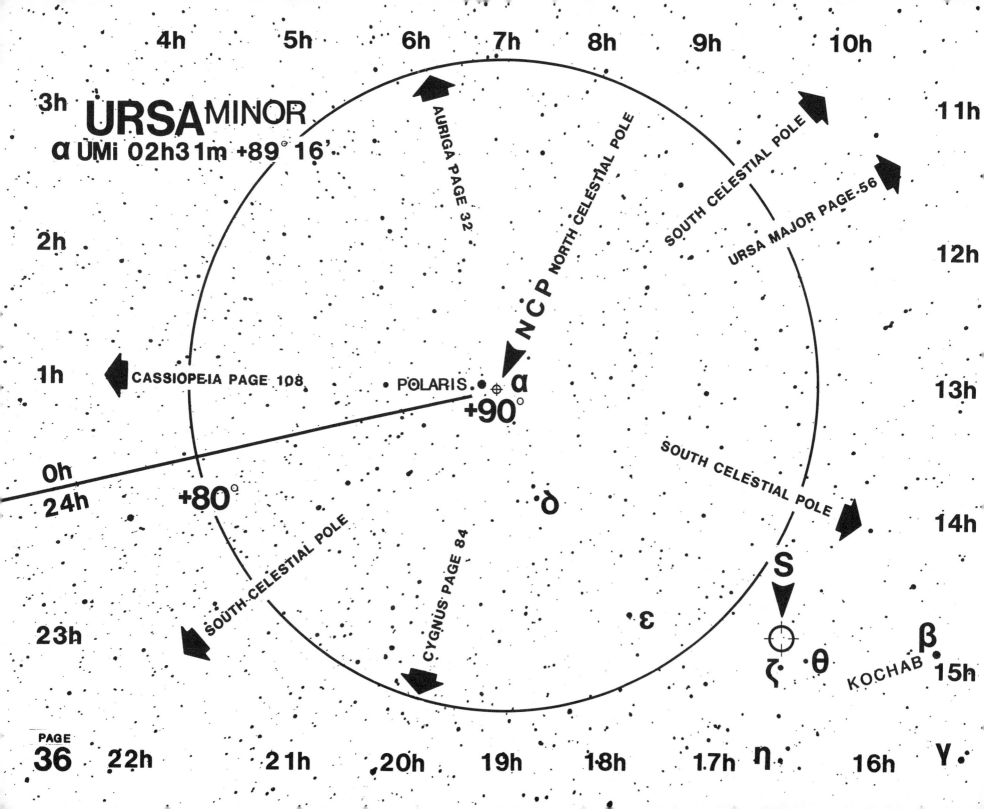

URSAMINOR

ALIGN STARFRAME WITH
POLARIS AND KOCHAB

POLARIS
(ALPHA)

N

KOCHAB
(BETA)

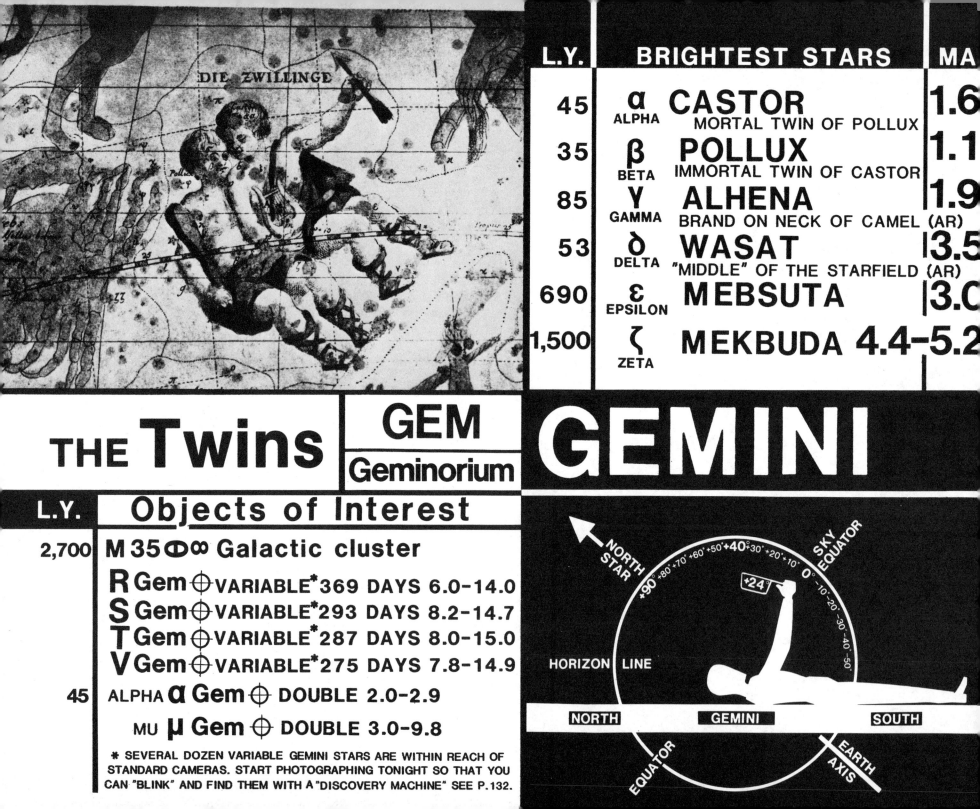

DIE ZWILLINGE

THE Twins

GEM
Geminorium

GEMINI

L.Y.	BRIGHTEST STARS	MA
45	α **CASTOR** ALPHA MORTAL TWIN OF POLLUX	1.6
35	β **POLLUX** BETA IMMORTAL TWIN OF CASTOR	1.1
85	γ **ALHENA** GAMMA BRAND ON NECK OF CAMEL (AR)	1.9
53	δ **WASAT** DELTA "MIDDLE" OF THE STARFIELD (AR)	3.5
690	ε **MEBSUTA** EPSILON	3.0
1,500	ζ **MEKBUDA** ZETA	4.4-5.2

L.Y.	Objects of Interest
2,700	**M 35** ⊕∞ Galactic cluster
	R Gem ⊕ VARIABLE*369 DAYS 6.0-14.0
	S Gem ⊕ VARIABLE*293 DAYS 8.2-14.7
	T Gem ⊕ VARIABLE*287 DAYS 8.0-15.0
	V Gem ⊕ VARIABLE*275 DAYS 7.8-14.9
45	ALPHA **α** Gem ⊕ DOUBLE 2.0-2.9
	MU **μ** Gem ⊕ DOUBLE 3.0-9.8

* SEVERAL DOZEN VARIABLE GEMINI STARS ARE WITHIN REACH OF STANDARD CAMERAS. START PHOTOGRAPHING TONIGHT SO THAT YOU CAN "BLINK" AND FIND THEM WITH A "DISCOVERY MACHINE" SEE P.132.

NORTH STAR

+90° +80° +70° +60° +50° **+40°** +30° +20° +10° 0° -10° -20° -30° -40° -50°

SKY EQUATOR

+24°

HORIZON LINE

NORTH   GEMINI   SOUTH

EQUATOR

EARTH AXIS

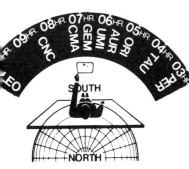

1 JAN. MIDNIGHT
15 JAN. 11 P.M.
1 FEB. 10 P.M.
15 FEB. 9 P.M.
1 MAR. 8 P.M.

Phone your nearest planetarium or observatory to find out about amateur astronomy groups, so that you can attend some of their winter meetings and decide whether you wish to join such a club in time for the spring and summer "star parties." You will meet interesting people and learn about the sky and telescopes.

RADIANT

CASTOR

POLLUX

GEMINI

February is a good time to start observing the constellation of Gemini, the heavenly Twins. Bright Castor lies at the head of what can be described as the northern "stick figure," and Pollux at the head of the southern. Gemini presents a recognizable pictogram.

Following Aries and Taurus, Gemini is the third constellation of the Zodiac. The sun passes here between June 21 and July 21.

Gemini and Canis Major (see the following constellation) lie in the same position in right ascension. The Twins' place in the sky is as far north of the celestial equator as the Great Dog lies south of it. Once you have learned to recognize these distinctive groupings, you may want to watch them go beyond their culminating positions and follow their westward journey until they "set" in the west. You will note that Gemini will remain visible longer and will set after the Great Dog. This is because our northern "flatforms" favor viewing northern constellations.

Alpha (α) Geminorium is a beautiful pair of double stars observable under good conditions with a telescope of 3 inches diameter. The stars in the pair are of magnitude 2 and 3 respectively. Each of the star pairs (called Castor A and Castor B) is composed of additional faint stars which are impossible to detect visually. We know today that together with yet another distant pair, named Castor C, Alpha (α) Gem consists of no fewer than six separate stars.

Pollux, although identified as Beta (β) Gem, is actually the brighter of the Twins. Just south of it lie the positions of the two variables S and T. You can patrol the constellation by photographing it at monthly intervals while it is in view. When you compare your pictures with future ones using a Problicom (pages 132-33), you are bound to find these variable beacons, together with others such as the variable R which is almost astride the ecliptic.

To the right lies M35. This binocular object is just visible to the naked eye under dark skies. Viewed with almost any telescope, it readily appears as a galactic cluster, and the brightest of approximately 150 stars can be distinguished. The position of the "radiant" of the December Geminid meteor shower is also marked (pages 125-26).

GALACTIC CLUSTER

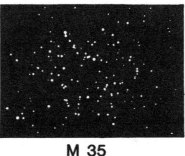

M 35

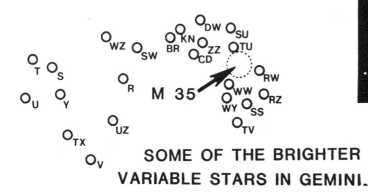

SOME OF THE BRIGHTER VARIABLE STARS IN GEMINI.

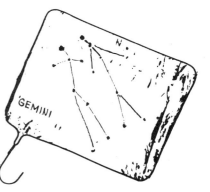

CANCER PAGE 48

7h

Find the pictogram stars on this page and connect them with ruled colored lines. You can also create your own ideograms with red markings which will filter out and disappear under a red flashlight (page 135).

•θ

TAURUS PAGE 24

⊕ ← RADIANT OF GEMINID METEORS (SEE PAGE 126)

CASTOR •α
ρ

+30°

•T

σ

POLLUX •β

ι

U

M 35

ε

T ►○ ○ ◄ S

R

ECLIPTIC

δ

μ η

◄ PATH OF THE SUN IN JUNE / JULY

ζ

ν

GEMINI
λ

β Gem 07h 45m +28°02'

γ

ORION PAGE 28

○ ◄ V

ξ

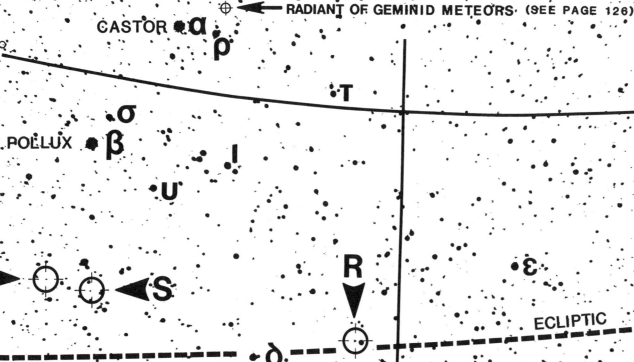

ALIGN WITH CASTOR AND POLLUX

N

STANDARD 50 MM LENS

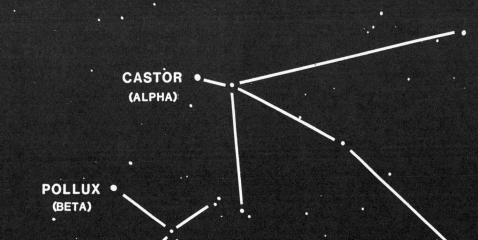

CASTOR
(ALPHA)

POLLUX
(BETA)

GEMINI

AIM CAMERA AT ⊕ TARGET BETWEEN TWINS

41

L.Y.	BRIGHTEST STARS		MA
8.7	α ALPHA	SIRIUS* "DOG STAR"	-1.5
750	β BETA	MIRZAM "ANNOUNCER" OF SIRIUS (AR)	2.0
1,250	γ GAMMA	MULIPHEIN	4.0
3,000	δ DELTA	WEZEN	1.9
490	ε EPSILON	ADHARA	1.5
2,500	η ETA	ALUDRA	2.4
290	ζ ZETA	PHURUD	3.0

WITH OMICRON O THE ARABS CALLED THESE FOUR STARS AL ADARA "THE VIRGINS"

* SIRIUS IS THE BRIGHTEST STAR IN THE HEAVENS

GREAT Dog

CMA
Canis Majoris

CANIS MAJOR

Objects of Interest

L.Y.	
2,400	M41 ⊕ Galactic cluster
	R CMa ⊕ VARIABLE* 1.1 DAYS 6.0-6.6
	α CMa ⊕ DOUBLE 1.5 & 8.6 (VERY DIFFICULT TO SEPARATE)
	EPSILON ε CMa ⊕ DOUBLE 1.6 & 8.1
	MU μ CMa ⊕ DOUBLE 4.6 & 9.5
	NU ν¹ CMa ⊕ DOUBLE 5.8 & 7.9

* THE CHANGES FOR SHORT-PERIOD VARIABLES ARE MEASURED IN DAYS OR HOURS. (PAGE 121)

NORTH STAR

+90° +80° +70° +60° +50° +40° +30° +20° +10° 0° -10° -20° -30° -40° -50°

SKY EQUATOR

22½°

HORIZON LINE

NORTH — CANIS MAJOR — SOUTH

EQUATOR

EARTH AXIS

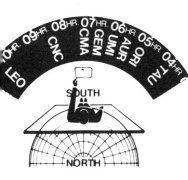

LEO 10HR. 09HR. 08HR. CNC 07HR. CMA 06HR. GEM 05HR. AUR UMI ORI TAU 04HR.

SOUTH

NORTH

1 JAN. MIDNIGHT
15 JAN. 11 P.M.
1 FEB. 10 P.M.
15 FEB. 9 P.M.
1 MAR. 8 P.M.

Last 1/4	New	First 1/4	Full
Jan. 29	Feb. 6	Feb. 12	Feb. 21

Let's get in step with the moon. Check your newspaper for the times of moonrise and moonset near the little moon images. Last- and first-quarter moons give you several hours of dark skies before or after midnight respectively. Near the new moon you will have good dark skies almost all night. Cut out and keep the little lunar almanacs. They also give useful sunrise and sunset times.

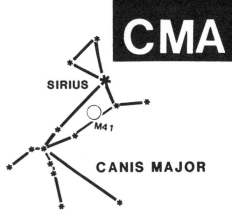

SIRIUS

M41

CANIS MAJOR

The constellation of the "Great Dog" lies just below Orion and a little to the left of it. Canis Major is also known as the Dog of Orion, and its constellation pictogram fairly represents a dog. With Sirius, the brightest star in the heavens on the dog collar, it should be easy to find.

In previous pages we have found that the brightest stars are of zero magnitude, and that as the magnitude numbers increase, the stars get fainter. Thus a star of magnitude 4 is less bright than one with a magnitude of 3, which in turn is fainter than a star of 2nd magnitude. First-magnitude stars are very bright. Sirius, Alpha (α) Canis Majoris, is so very dazzling that its magnitude is *minus* 1.5. Even among the planets, only Venus and Jupiter are brighter. Alpha (α) CMa is also called the "Dog Star." Yet another of its names, "Nile Star," refers to the time five thousand years ago when ancient Egyptians worshiped it. We know that its hieroglyphic was a dog. Each year when the Dog Star rose just before the hot morning sun and was visible for a little while at dawn, the priests knew that the Nile would soon overflow and spread its life-giving water on the waiting crops which had been planted on the riverbanks.

Isis and Osiris were gods related to Sirius. Giant temples were built for them, with their corridors oriented toward the point where the Dog Star would rise.

Sirius has a dwarf companion of magnitude 9 which can sometimes be seen with large telescopes. The mass of this tiny star is so compressed that a teaspoon of it weighs one ton.

Mirzam, Beta (β) CMa, was called the "Announcer" (of Sirius), because it runs ahead and rises just a little before its glamorous mate. You may want to verify this during the evening in December.

An open star cluster listed as M41 by Messier lies near the heart of the dog, some four degrees south of Sirius. It is believed that what we call M41 today was the faintest astronomical object ever recorded. Aristotle mentioned it in 325 B.C. as a "cloudy spot" in the sky. If you can, view this cluster through a telescope. You should be able to see as many as twenty stars out of a total of a hundred.

If you draw a straight line between the tail star Eta (η) CMa and Pollux in Gemini, you will find a very bright zero-magnitude star about two thirds of the way north. This is Procyon, the brightest star in Canis Minor.

Canis Minor, the Small Dog, has a pictogram consisting of a straight line, connecting Alpha (α) Canis Minoris with Beta (β) CMi, the next-brightest star lying four degrees to the right and above Procyon.

SIRIUS
HIEROGLYPH
2,000 B.C.

NORTH

1980 1985
1990
B
1995
SIRIUS A

ORBIT OF COMPANION OF SIRIUS

GALACTIC CLUSTER

M 41

EGYPTIAN TEMPLE

43

Find the pictogram stars on this page and connect them with ruled colored lines. You can also create your own ideograms with red markings which will filter out and disappear under a red flashlight (page 135).

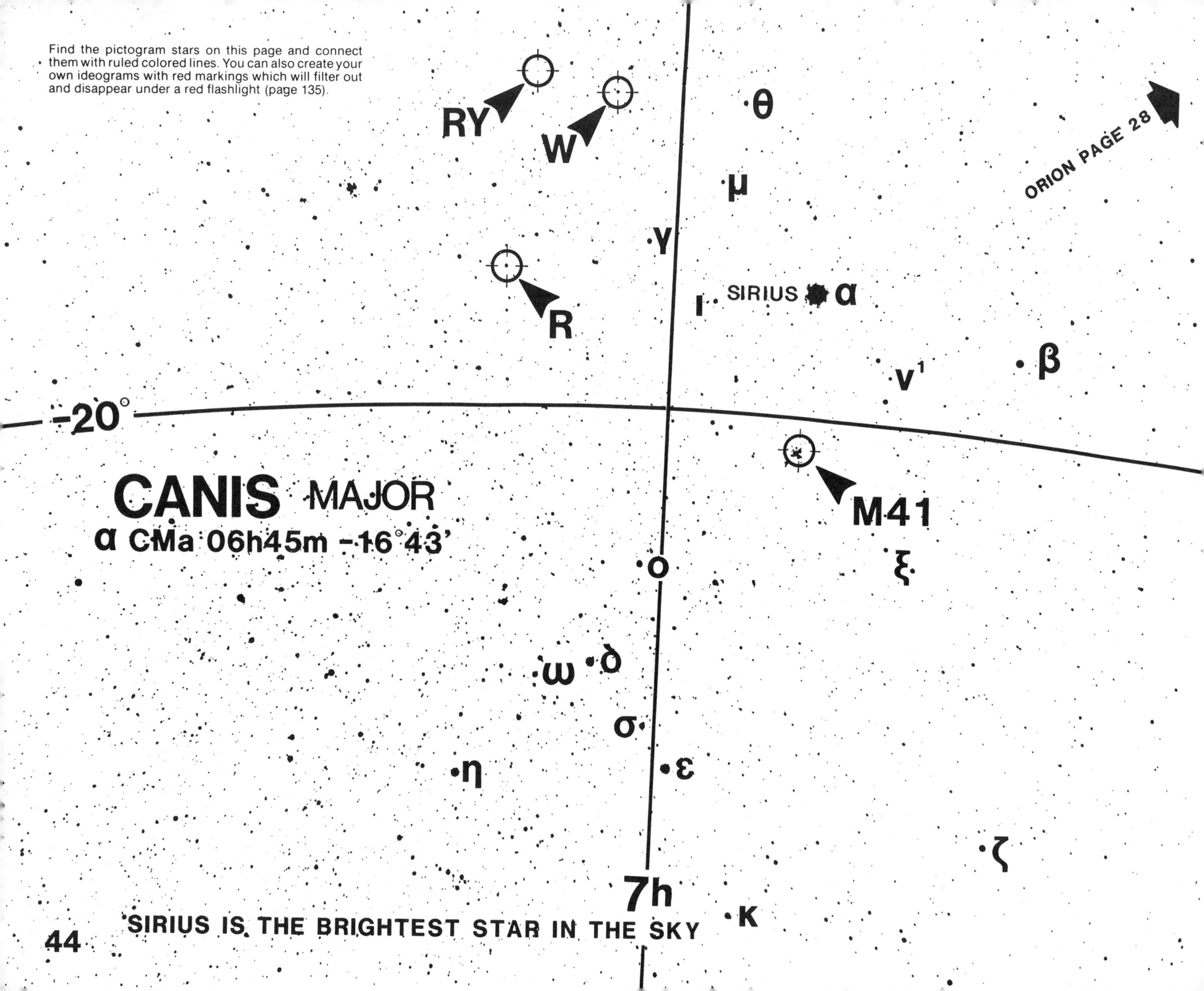

RY

W

θ

μ

γ

ORION PAGE 28

R

ι   SIRIUS ● α

ν¹   · β

−20°

CANIS MAJOR
α CMa 06h 45m −16°43'

M41

ξ

ο

ω ·δ

σ

ε

η

ζ

7h

κ

SIRIUS IS THE BRIGHTEST STAR IN THE SKY

44

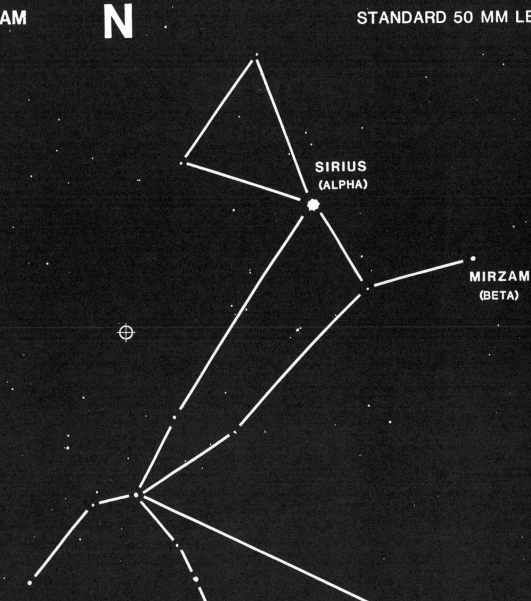

CANIS MAJOR

SIRIUS (ALPHA)

MIRZAM (BETA)

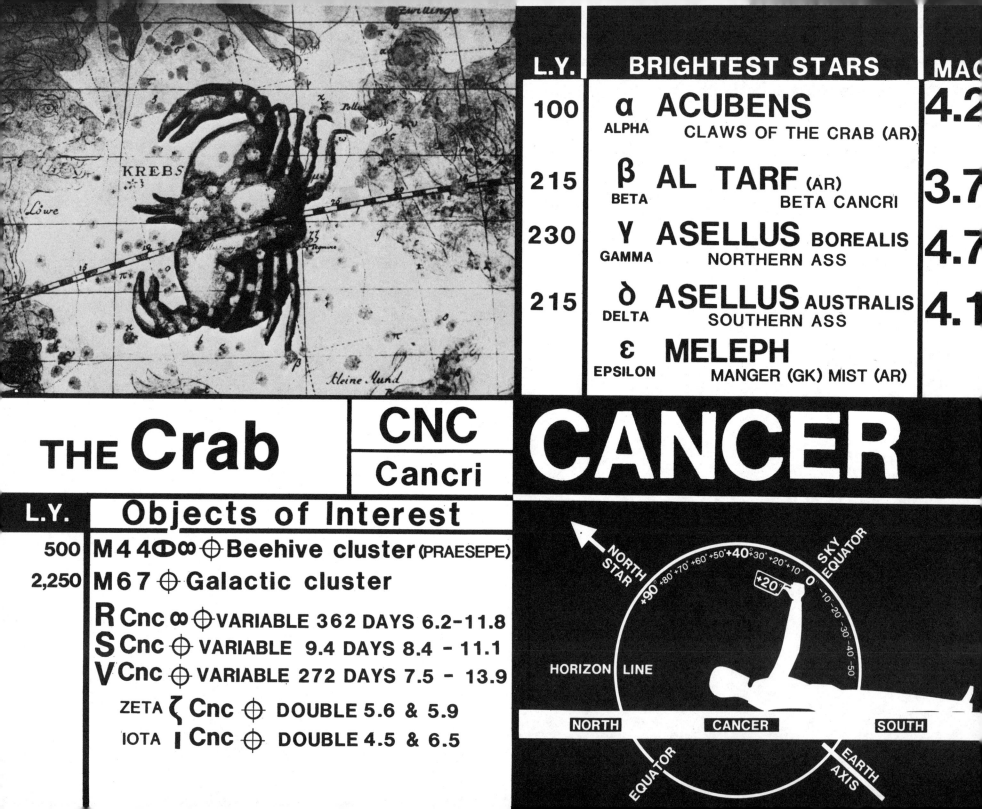

L.Y.	BRIGHTEST STARS	MAG
100	α ACUBENS — ALPHA — CLAWS OF THE CRAB (AR)	4.2
215	β AL TARF (AR) — BETA — BETA CANCRI	3.7
230	γ ASELLUS BOREALIS — GAMMA — NORTHERN ASS	4.7
215	δ ASELLUS AUSTRALIS — DELTA — SOUTHERN ASS	4.1
	ε MELEPH — EPSILON — MANGER (GK) MIST (AR)	

THE Crab

CNC / Cancri

CANCER

Objects of Interest

L.Y.	
500	M 44 ⊕∞⊕ Beehive cluster (PRAESEPE)
2,250	M 67 ⊕ Galactic cluster
	R Cnc ∞⊕ VARIABLE 362 DAYS 6.2-11.8
	S Cnc ⊕ VARIABLE 9.4 DAYS 8.4 - 11.1
	V Cnc ⊕ VARIABLE 272 DAYS 7.5 - 13.9
	ZETA ζ Cnc ⊕ DOUBLE 5.6 & 5.9
	IOTA ι Cnc ⊕ DOUBLE 4.5 & 6.5

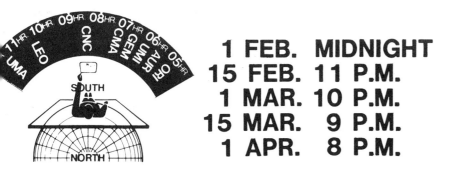
1 FEB. MIDNIGHT
15 FEB. 11 P.M.
1 MAR. 10 P.M.
15 MAR. 9 P.M.
1 APR. 8 P.M.

Prepare to make and trace your spring, summer and autumn star-frames now, while it is still chilly. Then you will be ready, when the warm nights come, to set out on your journey to the stars.

The constellation of Cancer is not easy to find. None of its stars is brighter than Beta (β) Cancri, the brightest in the starfield at magnitude 3.7. The star Alpha (α) Cnc at magnitude 4.2 is half a magnitude fainter. It bears the name "Acubens," referring to scissors or claws. The ancient Egyptians regarded the creature as a scarab, the industrious beetle they revered.

Cancer is the least prominent of the twelve constellations of the Zodiac. The sun moves through the field from July 21 until August 11, with the ecliptic cutting close by Delta (δ) Cancri, the hub of the constellation pictogram.

A famous open star cluster is situated just north and a little to the right of Delta (δ) Cancri. It is called Praesepe, or the Beehive, and is one of the brightest and largest galactic clusters near us. The object was mentioned in astronomical literature over two thousand years ago as a "little cloud" or "patch of mist." It remained a mystery until the telescope was invented.

In his 1610 publication *The Starry Messenger,* Galileo made a drawing of what he had seen through his new telescope and described that the nebulous Praesepe patch is a mass of nearly forty little stars. Today we know that the Beehive consists of about 350 stars ranging in magnitude from 6, the limit of what can be seen with the naked eye under dark skies, to telescopic magnitude 17. Messier gave the number 44 to the sparkling cluster. Another fainter such grouping, called M67, is just to the right of Acubens. Here too about five hundred stars are grouped together.

It is said that Praesepe was used as a weather indicator in ancient times and that the clarity with which it could be seen foretold if storms lay ahead. Today it offers a useful visual scale by which we can judge the degree of transparency for sky observations.

Two of the brighter variable stars can be observed periodically in Cancer. R Cancri goes through its entire cycle from faintest "minimum," at magnitude 12, to brightest "maximum" at magnitude 6 and back again over a period of 362 days. The variable V Cnc, with a magnitude range from 7.5 to 14, takes 272 days to go through its paces. The forces which are at work to bring about such radical but constant changes in the makeup of stars are phenomenal. The constellation of Cancer has relatively few variables — only about eighty (see pages 121-22).

M 44 BEEHIVE (PRAESEPE)

GALACTIC CLUSTER

M 67

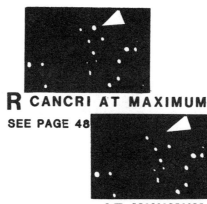

R CANCRI AT MAXIMUM

SEE PAGE 48

AT MINIMUM

SCARAB

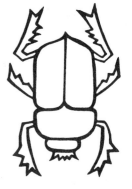

47

CASTOR

Find the pictogram stars on this page and connect them with ruled colored lines. You can also create your own ideograms with red markings which will filter out and disappear under a red flashlight (page 135).

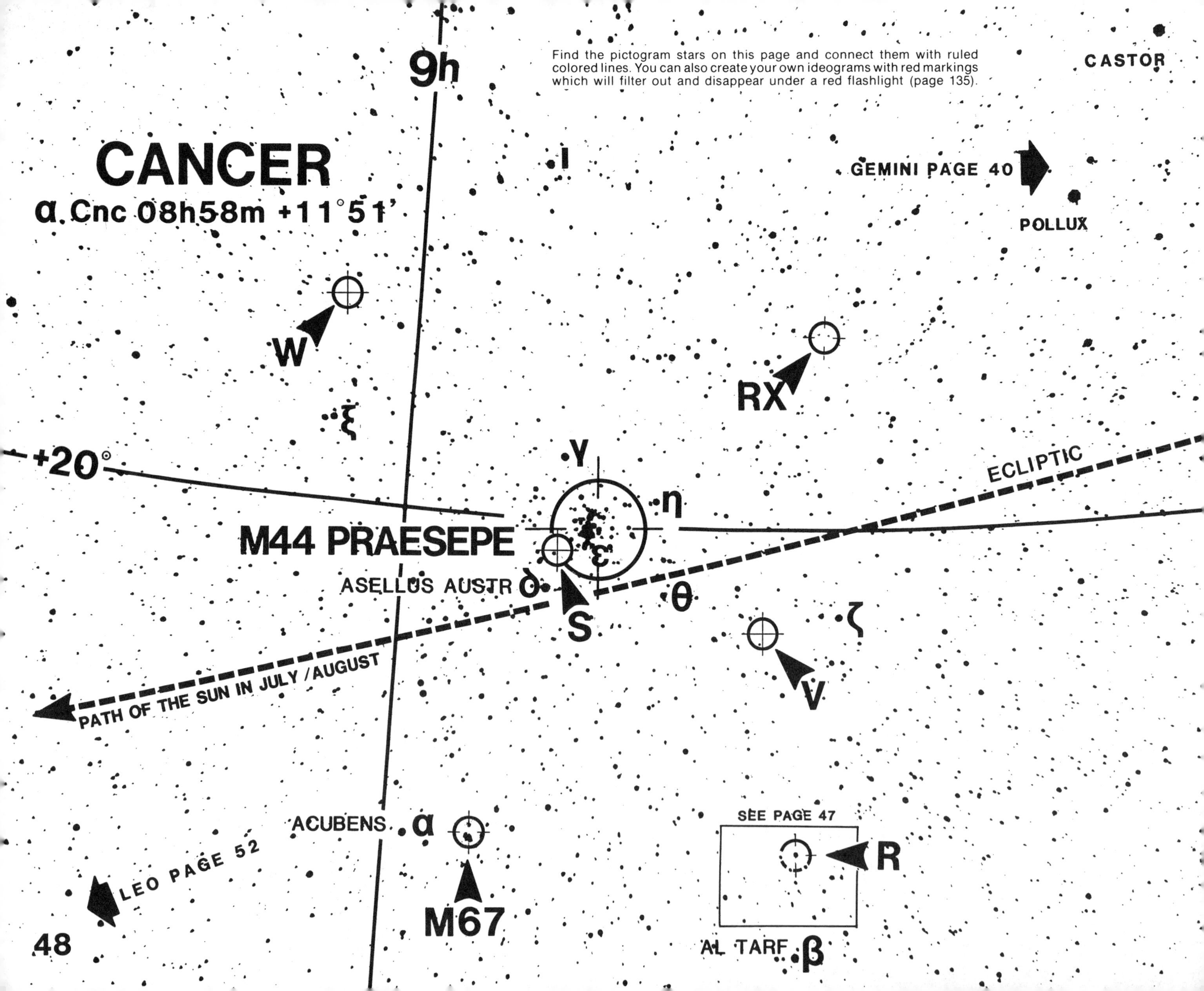

CANCER
α. Cnc 08h58m +11°51'

I

GEMINI PAGE 40 →

POLLUX

W

RX

ξ

γ

η

+20°

ECLIPTIC

M44 PRAESEPE

ε

ASELLUS AUSTR δ

θ

S

ζ

ν

PATH OF THE SUN IN JULY /AUGUST ←

ACUBENS α

SEE PAGE 47

R

LEO PAGE 52 ←

M67

AL TARF β

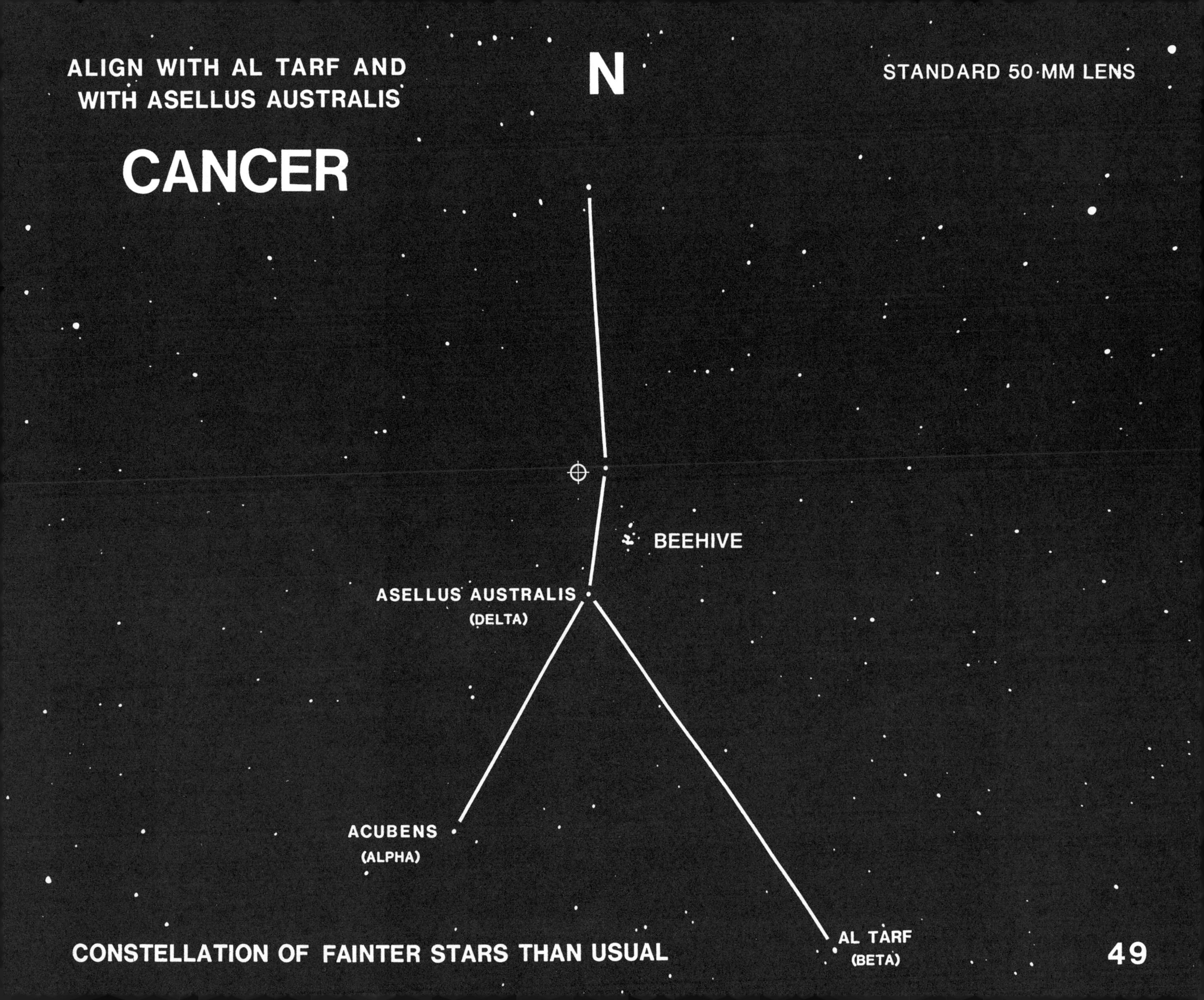

ALIGN WITH AL TARF AND
WITH ASELLUS AUSTRALIS

N

STANDARD 50·MM LENS

CANCER

BEEHIVE

ASELLUS AUSTRALIS
(DELTA)

ACUBENS
(ALPHA)

AL TARF
(BETA)

CONSTELLATION OF FAINTER STARS THAN USUAL

49

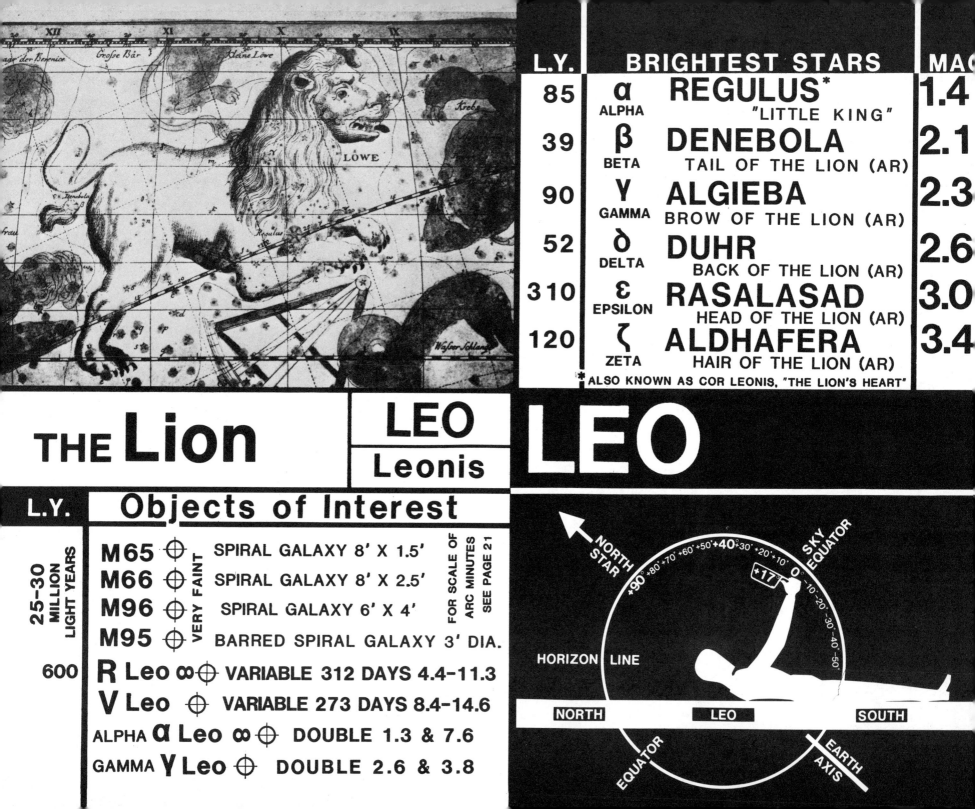

THE Lion

LEO
Leonis

LEO

L.Y.	BRIGHTEST STARS	MAG
85	α ALPHA **REGULUS*** "LITTLE KING"	1.4
39	β BETA **DENEBOLA** TAIL OF THE LION (AR)	2.1
90	γ GAMMA **ALGIEBA** BROW OF THE LION (AR)	2.3
52	δ DELTA **DUHR** BACK OF THE LION (AR)	2.6
310	ε EPSILON **RASALASAD** HEAD OF THE LION (AR)	3.0
120	ζ ZETA **ALDHAFERA** HAIR OF THE LION (AR)	3.4

*ALSO KNOWN AS COR LEONIS, "THE LION'S HEART"

Objects of Interest

L.Y.			
25-30 MILLION LIGHT YEARS	M65 ⊕	VERY FAINT	SPIRAL GALAXY 8' X 1.5'
	M66 ⊕		SPIRAL GALAXY 8' X 2.5'
	M96 ⊕		SPIRAL GALAXY 6' X 4'
	M95 ⊕		BARRED SPIRAL GALAXY 3' DIA.
600	R Leo ∞ ⊕		VARIABLE 312 DAYS 4.4-11.3
	V Leo ⊕		VARIABLE 273 DAYS 8.4-14.6
	ALPHA α Leo ∞ ⊕		DOUBLE 1.3 & 7.6
	GAMMA γ Leo ⊕		DOUBLE 2.6 & 3.8

FOR SCALE OF ARC MINUTES SEE PAGE 21

NORTH STAR

+90° +80° +70° +60° +50° +40° +30° +20° +10° 0° -10° -20° -30° -40° -50°

+17°

SKY EQUATOR

HORIZON LINE

| NORTH | LEO | SOUTH |

EQUATOR

EARTH AXIS

Did you miss cold January or February constellations? It is not too late to catch them. Look for them soon after sundown, when they will still culminate in their highest position. From now on try to "follow them down" in the evenings until they set.

1 MAR.   MIDNIGHT
15 MAR.   11 P.M.
1 APR.   10 P.M.
15 APR.   9 P.M.
1 MAY   8 P.M.

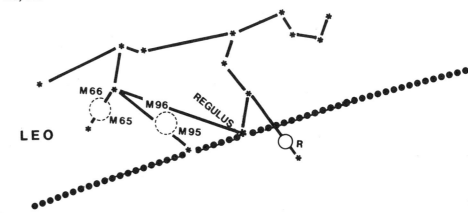

LEO

Throughout antiquity the constellation of the Lion has been associated with the sun, and Regulus, its brightest star, with royalty. The name "Regulus" means Little King. Alpha (α) Leo was often referred to as "Cor Leonis," the Lion's Heart. Thus Leo has always held an exalted position among the stars, which extended to the Zodiac. The sun is in Leo from August 12 to September 16.

A four-legged animal with head and tail can be readily recognized, and the characteristic mane of the lion may be found in the part of the constellation which some prefer to identify as a sickle.

Alpha (α) Leonis, the twenty-first brightest star in the heavens at magnitude 1.4, lies at the base of the sickle. It can also be said to mark the lion-paw on one of the beast's forelegs. Some have gone so far as to refer to the sickle as a backward question mark, where Regulus becomes the "dot."

Just like Delta (δ) Cancri, Alpha (α) Leonis lies almost astride the ecliptic. You can picture the position of the path of the sun if you draw an arc from the point of today's sunrise through Regulus back down the horizon where the sun set this evening. The position of the moon, which travels in a closely matching orbit, will be very close to the ecliptic and will confirm your estimate. Bright planets, when they are in Leo, will help you visualize the width of the Band of the Zodiac, which extends from 8 degrees above to 8 degrees below Regulus.

Both Alpha (α) Leonis and Gamma (γ) Leonis are double stars. Of the two, Algieba may be the more difficult to observe because the pair is separated by only 4 arc-seconds.* In the case of Regulus, the separation is nearly 3 arc-minutes† (page 21).

The variable R Leo, close to the paw on the foremost leg, is a naked-eye object while at maximum and may be observable with binoculars even from cities when it is at its brightest. Some astronomical publications announce major variable star maximums, so you can be on the lookout for such events. R Leonis has a period of 312 days and fades to magnitude 11.3 at its minimum.

M65, M66, M95 and M96 are faint spiral galaxies for which, like Messier, you will need a telescope (page 118).

*1° (degree) = 60' (arc-minutes).
†1' (arc-minute) = 60" (arc-seconds).

THE SICKLE.

REGULUS

SPIRAL GALAXY

M 65   8' X 1.5'

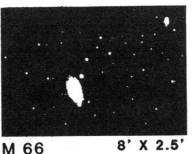

SPIRAL GALAXY

M 66   8' X 2.5'

BARRED SPIRAL GALAXY

M 95   3' DIAMETER

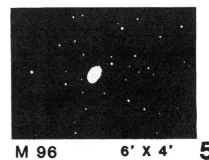

SPIRAL GALAXY

M 96   6' X 4'

51

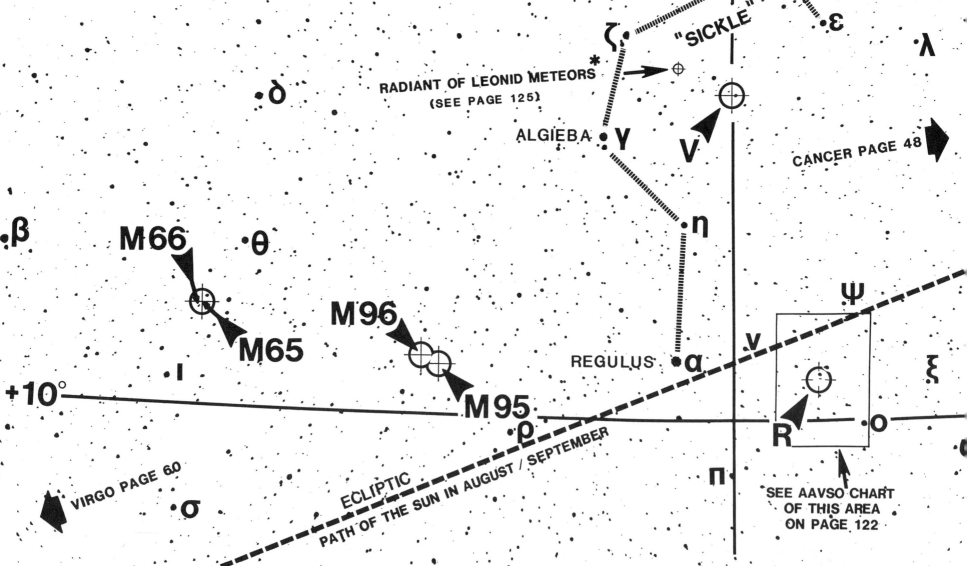

Find the pictogram stars on this page and connect them with ruled colored lines. You can also create your own ideograms with red markings which will filter out and disappear under a red flashlight (page 135).

10h

κ

μ

ε

λ

LEO
α Leo 10h08m +11°58'

ζ

"SICKLE"

δ

RADIANT OF LEONID METEORS
(SEE PAGE 125)

*

⊕

⊕

CANCER PAGE 48

ALGIEBA · γ

ν

β

M66

θ

η

M65

M96

ψ

M95

ν

REGULUS · α

ξ

ρ

R

ο

+10°

ι

π

SEE AAVSO CHART
OF THIS AREA
ON PAGE 122

VIRGO PAGE 60

σ

ECLIPTIC
PATH OF THE SUN IN AUGUST / SEPTEMBER

52 *METEORS OBSERVED BETWEEN NOVEMBER 14-20 WHICH SEEM TO ORIGINATE
FROM HERE ARE PART OF THE LEONID METEOR SHOWER. (SEE PAGES 125-126).

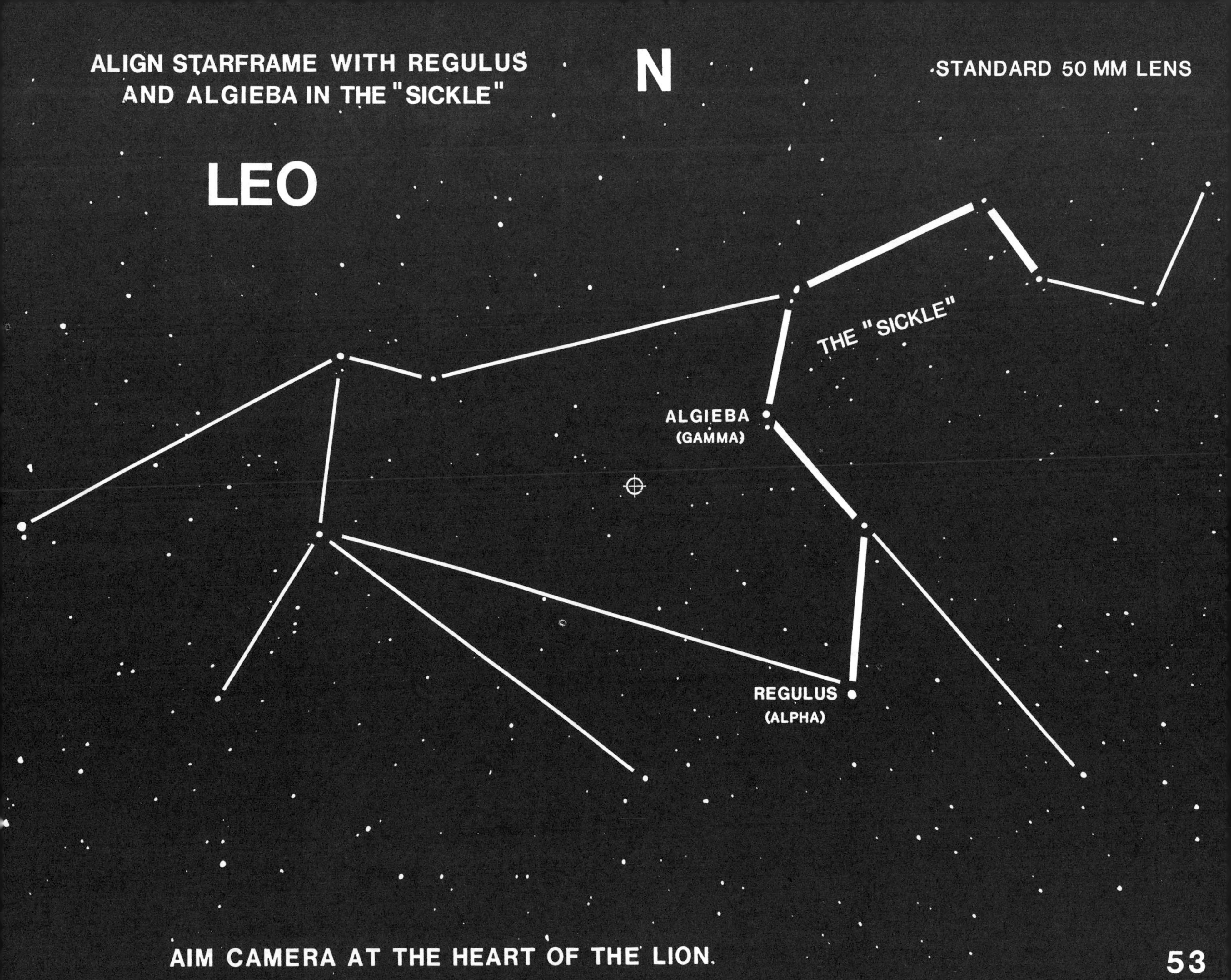

ALIGN STARFRAME WITH REGULUS
AND ALGIEBA IN THE "SICKLE"

N

STANDARD 50 MM LENS

LEO

THE "SICKLE"

ALGIEBA
(GAMMA)

REGULUS
(ALPHA)

AIM CAMERA AT THE HEART OF THE LION.

53

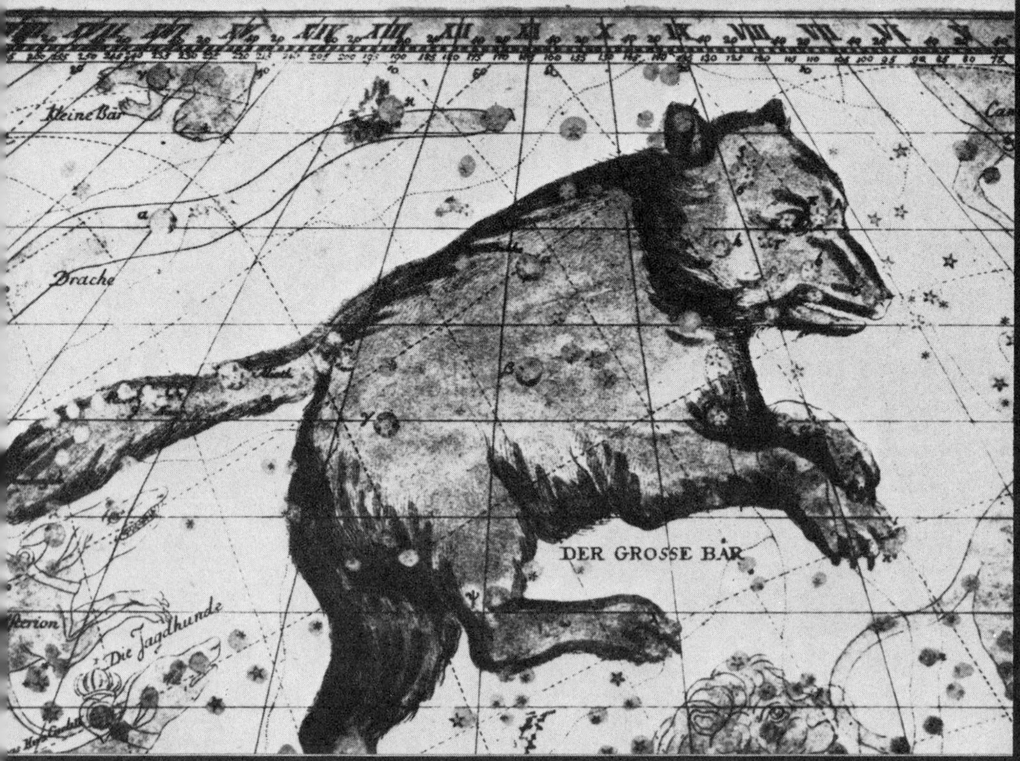

Drache
DER GROSSE BAR
Die Jagdhunde

L.Y.		BRIGHTEST STARS	MAG
75	α ALPHA	DUBHE — BEAR (AR)	1.8
62	β BETA	MERAK — FLANK OF THE BEAR (AR)	2.3
75	γ GAMMA	PHECDA — THIGH (AR)	2.4
65	δ DELTA	MEGREZ — ROOT OF THE TAIL (AR)	3.3
62	ε EPSILON	ALIOTH	1.7
59	ζ ZETA	MIZAR (ALCOR) (AND RIDER) — THE HORSE	2.2
110	η ETA	ALKAID — LEADER OF MOURNING DAUGHTERS (AR)	1.8

GREAT Bear

UMA

Ursae Majoris

URSA MAJOR

L.Y.	Objects of Interest
7 MILLION	M81 ⊕ FAINT SPIRAL GALAXY 21′ X 10′
	M82 ⊕ FAINT IRREGULAR GALAXY 9′ X 4′
	IN LOW-POWER TELESCOPES M81 AND M82 MAY BE VIEWED IN THE SAME FIELD, 35′ APART.
1,350	RUMa ⊕ VARIABLE 301 DAYS 6.7–13.4
	SUMa ⊕ VARIABLE 226 DAYS 7.0–12.4
	TUMa ⊕ VARIABLE 256 DAYS 6.6–13.4
	ZETA ζ UMa ⊕ DOUBLE 2.2 & 3.9
	XI ξ UMa ⊕ DOUBLE 3.7 & 9.7

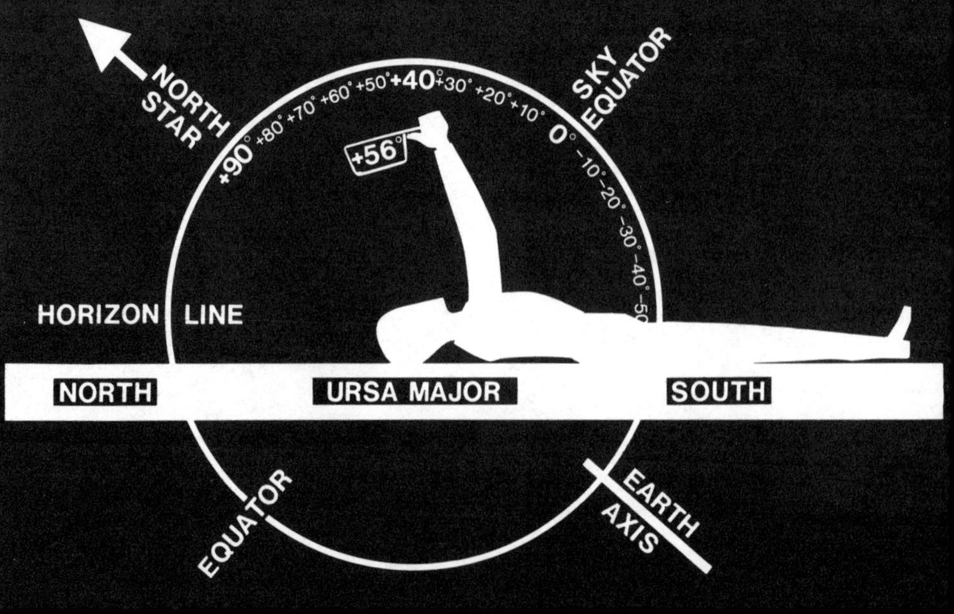

NORTH STAR
SKY EQUATOR
+90° +80° +70° +60° +50° +40° +30° +20° +10° 0° -10° -20° -30° -40° -50
+56°
HORIZON LINE
NORTH
URSA MAJOR
SOUTH
EQUATOR
EARTH AXIS

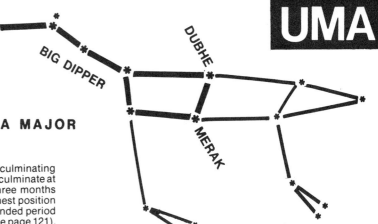

For DAYLIGHT SAVING TIMES see note on page 10.

1 MAR. MIDNIGHT
15 MAR. 11 P.M.
1 APR. 10 P.M.
15 APR. 9 P.M.
1 MAY 8 P.M.

URSA MAJOR

You may wish to photograph the constellation culminating at 4 A.M. tomorrow. One month from now it will culminate at 2 A.M. and at midnight two months hence. Three months from now it can still be photographed in its highest position at 10 P.M. With such records taken over an extended period of time you can keep an eye on variable stars (see page 121).

It is fascinating to learn that the constellation of the familiar seven stars in the northern skies which have been known as the Great Bear through Greek, Roman and Arab history were associated with the same animal by early Indian tribes of North America. It is doubly puzzling because the starfield of the Big Dipper bears no resemblance to a grizzly.

Tracing the origin of the name "Dubhe," that of the brightest star, seems easier. In contemporary Hebrew, just as in the Biblical form, the name for "bear" is Dovh. As in so many other cases, the name of the Alpha (α) star is often used to identify the entire starfield. Thus the related Arabic name *Al Dubh al Akbar* quite literally translates into "the Greater Bear."

Brightest Dubhe and the nearby star Merak together have long been called the "pointer stars." The distance between them, extended five times, leads to Polaris, as shown on the right. Using the Big Dipper as an aid to orientation will allow you to find the direction of the North Celestial Pole and the North Star during the first five or six months of each year. If you live in a far-northern latitude, Ursa Major will never set and you can use it for a compass all year.

As the star diagram indicates, the Big Dipper (also called the "Plow" or the "Wagon" in some parts of Europe) forms only part of the constellation. Different pictograms have been drawn over the centuries. All involve fainter stars and are difficult to identify. The diagram shown on page 57 includes a head and four legs to join with the dipper portion of the rump and tail. These will give us our bearings to find the rest of the animal.

The star Mizar, Zeta (ζ) Ursae Majoris, at magnitude 2.2 has a fairly bright naked-eye 4th-magnitude companion called Alcor. Together they are known as "the Horse and the Rider." They are said to present a test for eyesight. Binoculars will easily and clearly separate the pair into its two components.

M81 and M82 are objects listed in Messier's catalogue but are credited to his contemporary J. E. Bode, author of the star atlas from which the historical maps opposite were drawn. Bode discovered these two galaxies on the night of December 31 in 1774. M81 is a spiral galaxy not unlike our own Milky Way. M82 is an "irregular" galaxy, showing traces of an explosive past. In most telescopes the two objects can be observed in the same field of view. Both are about 7 million light-years away from us.

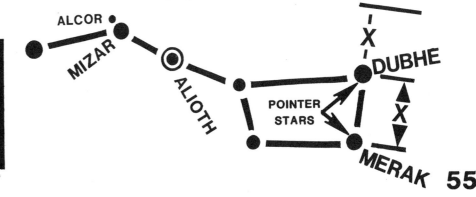

SPIRAL GALAXY

M 81   21' X 10'

IRREGULAR GALAXY

M 82   9' X 4'

55

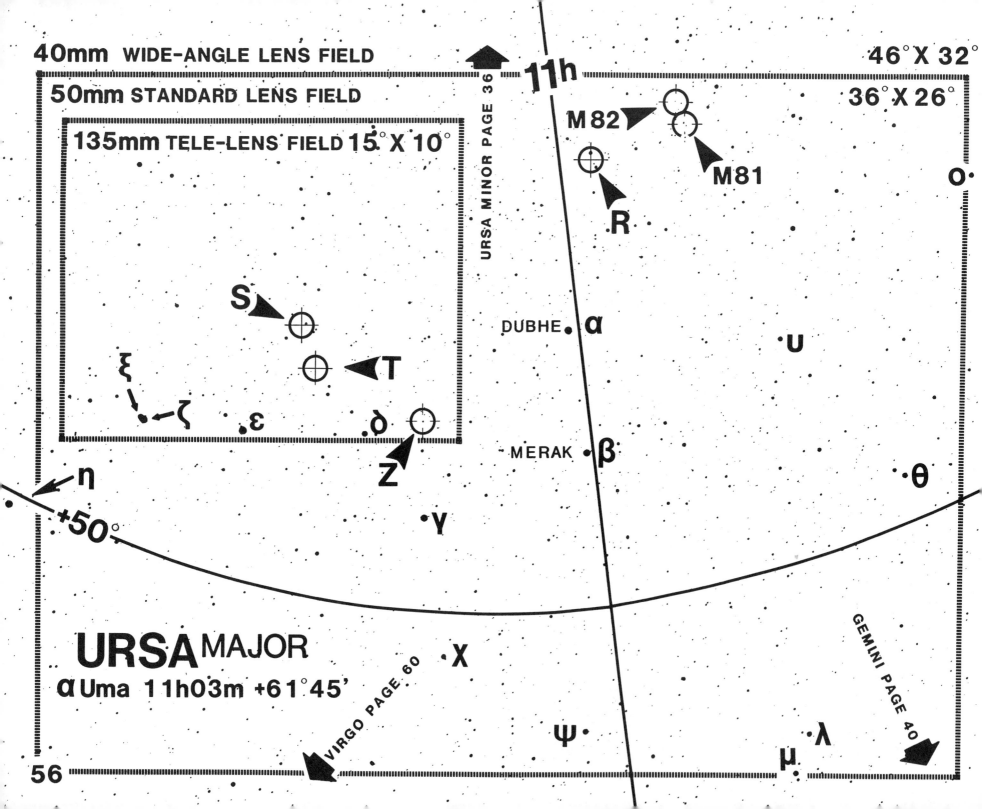

ALIGN WITH DUBHE AND MERAK

N

35 mm WIDE-ANGLE LENS FIELD
52° X 35°

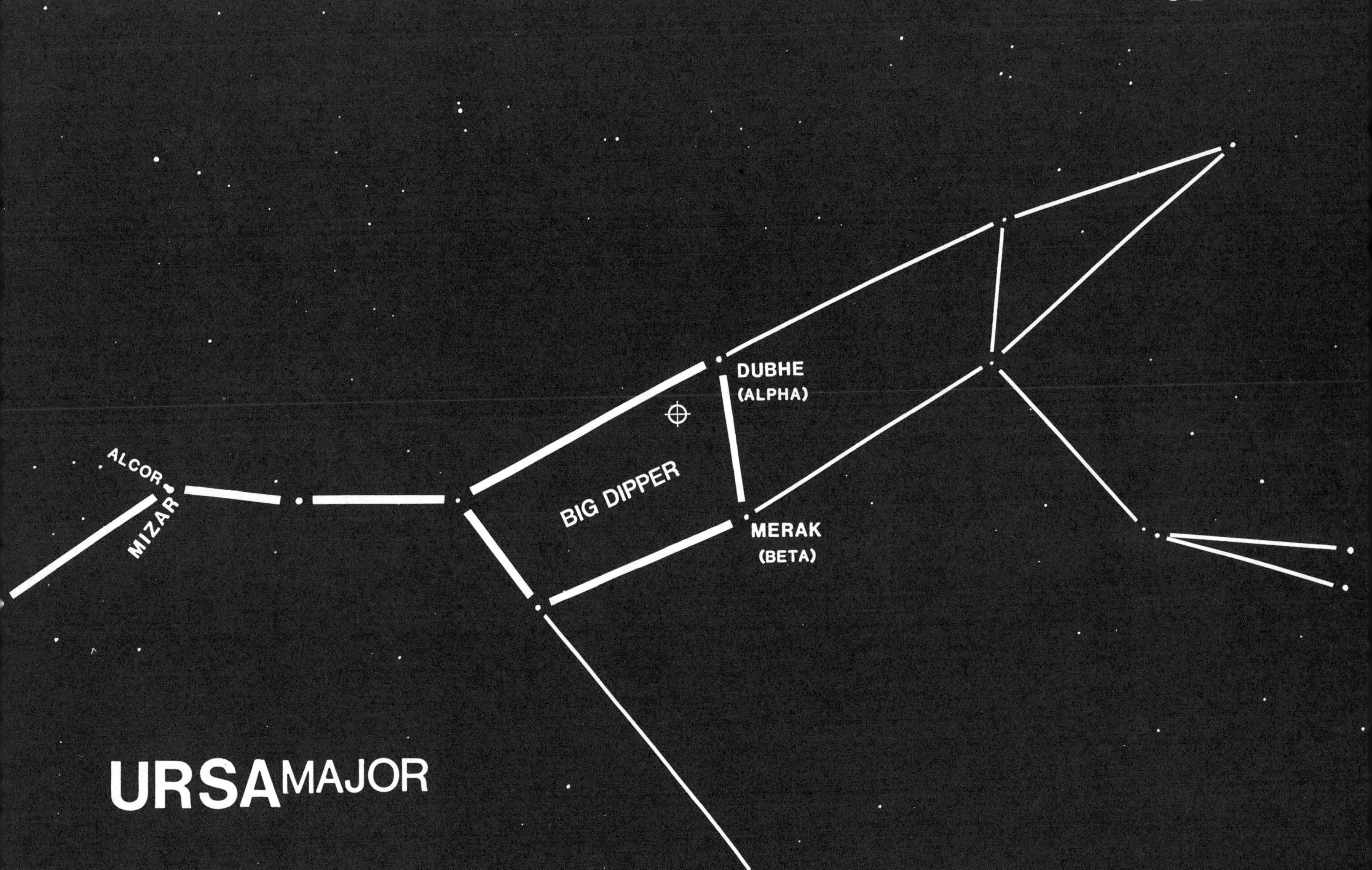
DUBHE
(ALPHA)
MERAK
(BETA)
BIG DIPPER
ALCOR
MIZAR
URSA MAJOR

IN WIDE ANGLE PHOTOGRAPHS
IMAGES AND SPACING WILL APPEAR SMALLER
TO COMPENSATE HOLD STARFRAME SLIGHTLY CLOSER TO EYES

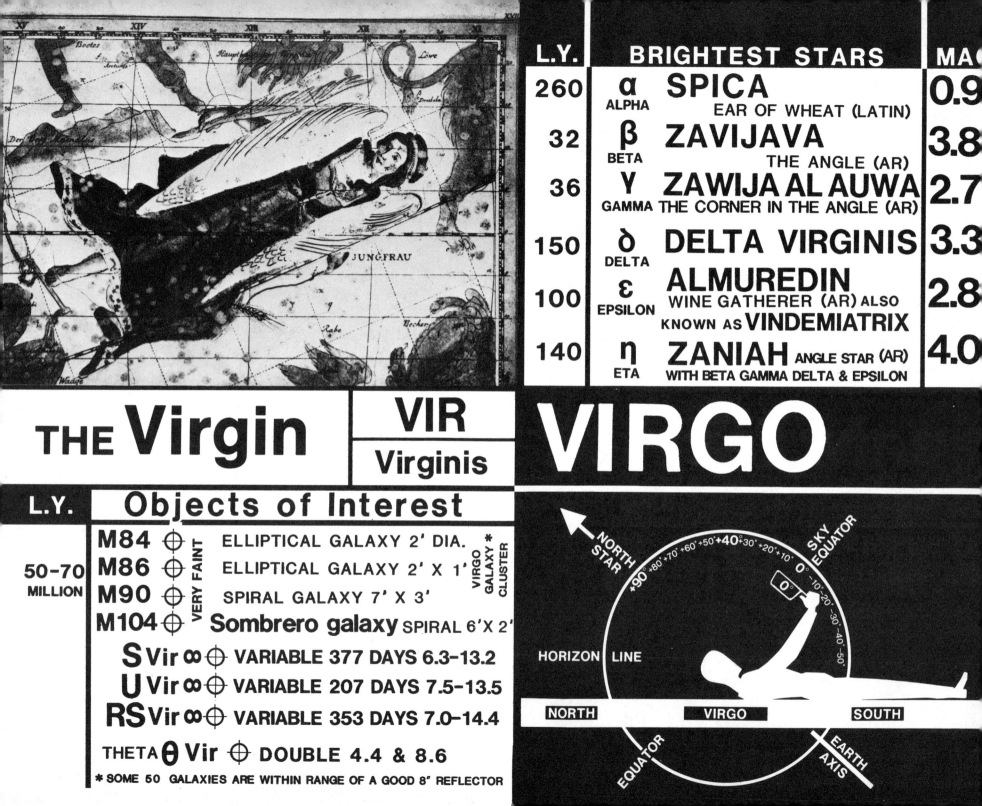

L.Y.	BRIGHTEST STARS	MAG
260	α ALPHA **SPICA** EAR OF WHEAT (LATIN)	0.9
32	β BETA **ZAVIJAVA** THE ANGLE (AR)	3.8
36	γ GAMMA **ZAWIJA AL AUWA** THE CORNER IN THE ANGLE (AR)	2.7
150	δ DELTA **DELTA VIRGINIS**	3.3
100	ε EPSILON **ALMUREDIN** WINE GATHERER (AR) ALSO KNOWN AS **VINDEMIATRIX**	2.8
140	η ETA **ZANIAH** ANGLE STAR (AR) WITH BETA GAMMA DELTA & EPSILON	4.0

THE **Virgin** | **VIR** Virginis | **VIRGO**

Objects of Interest

L.Y.		
50-70 MILLION	M84 ⊕	ELLIPTICAL GALAXY 2' DIA.
	M86 ⊕	ELLIPTICAL GALAXY 2' X 1'
	M90 ⊕	SPIRAL GALAXY 7' X 3'
	M104 ⊕	**Sombrero galaxy** SPIRAL 6' X 2'

VERY FAINT — VIRGO GALAXY CLUSTER

S Vir ∞ ⊕ VARIABLE 377 DAYS 6.3-13.2
U Vir ∞ ⊕ VARIABLE 207 DAYS 7.5-13.5
RS Vir ∞ ⊕ VARIABLE 353 DAYS 7.0-14.4

THETA **θ** Vir ⊕ DOUBLE 4.4 & 8.6

* SOME 50 GALAXIES ARE WITHIN RANGE OF A GOOD 8" REFLECTOR

NORTH STAR

+90° +80° +70° +60° +50° +40° +30° +20° +10° 0° -10° -20° -30° -40° -50°

SKY EQUATOR

0°

HORIZON LINE

| NORTH | VIRGO | SOUTH |

EQUATOR   EARTH AXIS

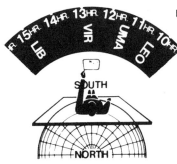

VIRGO CLUSTER

15 APR. MIDNIGHT
1 MAY 11 P.M.
15 MAY 10 P.M.
1 JUN. 9 P.M.
15 JUN. 8 P.M.

Many universities have departments of astronomy where someone can guide you toward amateur astronomy clubs or groups. Phone around today to plan your future activities. You will enjoy sharing your interest in the stars with others, and will find amateur astronomers in all walks of life.

VIRGO

SPICA

The Zodiac constellation of Virgo reclines in the night sky with her head pointed to the right, legs extending to the left. As shown in Bode's free interpretation of the figure, the virgin holds a spica (an ear of wheat), in her left hand and a palm branch in her right. Both are long-ago symbols of harvest time, when the sun passed through Virgo. Today the sun journeys through this constellation between September 21 and November 1 and crosses here from the Northern to the Southern Hemisphere at the time of the Autumn Equinox (pages 140-41).

Spica is a very bright star just south of the ecliptic. If you draw an imaginary line between Regulus, in adjoining Leo, and Spica, it will define the path of the sun. Again, the moon and any planets which may be in the area will all be traveling along or very close to this imaginary ecliptic.

Alpha (**α**) Virginis has a magnitude of 0.9 but fluctuates very slightly above and below this value. Variables with easy-to-detect cycles include the more prominent stars S Vir, U Vir and RS Vir, all of which may be observed with binoculars as they go through part of their changes.

Gamma (**γ**) Virginis is a double star with two components of almost identical 3rd magnitudes. Their motions bring them near each other every 171 years to where only the largest telescopes can separate them. They will be increasingly difficult to observe as we draw nearer to the end of our century.

Four Messier objects in Virgo are illustrated below. They are M84, M86, M90 and M104. In his description of the first three, Messier wrote: *"Nébuleuse sans étoile,"* nebula without stars. Little could he know that the hazy patches were galaxies and that each galaxy was, in fact, filled with millions of stars. More yet, in the area where these relatively bright objects were located, at least three thousand more galaxies have since been observed and identified, each one an island universe unto itself (page 118). With an 8-inch-diameter telescope you can see fifty or more of these distant worlds.

COMA BERENICES

VIRGO

CLUSTER OF GALAXIES

SOMBRERO GALAXY

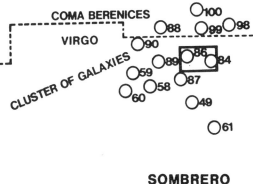

SOMBRERO GALAXY
◯ M104

M 86 2' X 1'   M 84 2' DIA.

M 90   7' X 3'

M 104   6' X 2'

59

Find the pictogram stars on this page and connect them with ruled colored lines. You can also create your own ideograms with red markings which will filter out and disappear under a red flashlight (page 135).

HERCULES PAGE 72

M 84

VIRGO
GALAXY
CLUSTER

M 90

M 86

o

VIRGO
αVir 13h25m -11°09'

ε

ν

RS

R

δ

U

SS

β

T

O°

SKY (CELESTIAL) EQUATOR

ζ

η

ZAWIJA AL AUWA • γ

POINT OF AUTUMN
EQUINOX
SEE PAGES 140–141

θ

PATH OF THE SUN IN OCTOBER

μ

I

S

LIBRA PAGE 64

M 104

LEO PAGE 52

K

ECLIPTIC

SPICA • α

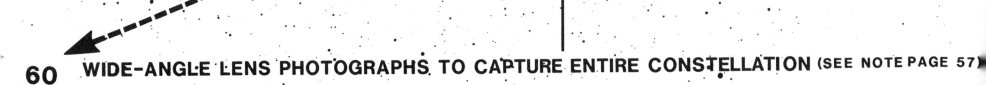

VIRGO

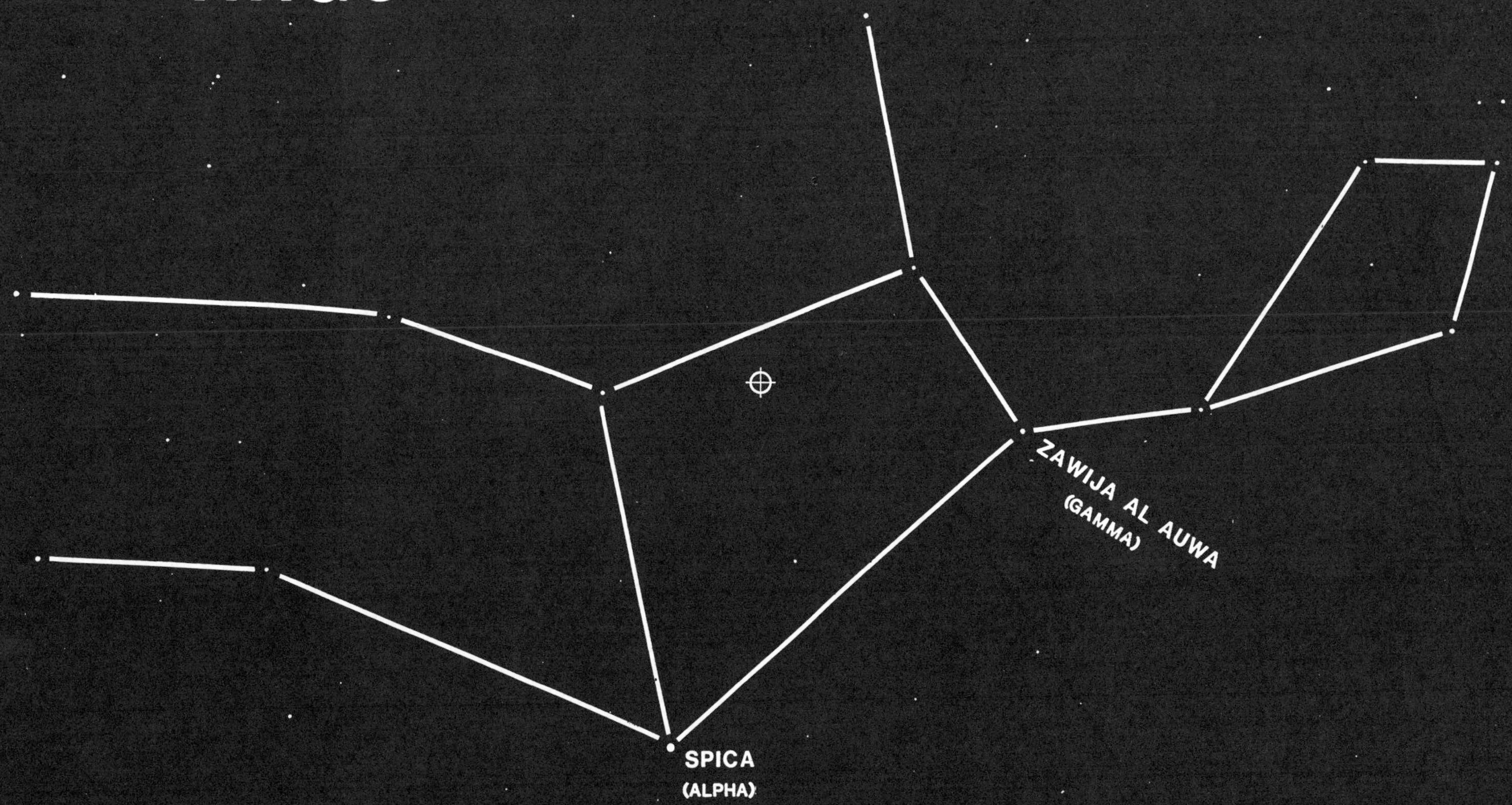

ZAWIJA AL AUWA
(GAMMA)

SPICA
(ALPHA)

BRIGHTEST STARS

L.Y.		MAG
72	α ALPHA **ZUBEN ELGENUBI** RIGHT CLAW OF THE SCORPION (AR) THE SOUTHERN CLAW	2.7
120	β BETA **ZUBEN ELSCHAMALI** LEFT CLAW OF THE SCORPION (AR) THE NORTHERN CLAW	2.6
109	γ GAMMA **ZUBEN HAKRABI** * TWO CLAWS OF THE SCORPION (AR)	4.0

* LIBRA STARS WERE ONCE PART OF SCORPIUS

THE Scales

LIB
Librae

LIBRA

Objects of Interest

USE CAMERA TO FIND AND TELESCOPE ⊕ TO VIEW.

L.Y.		
200	DELTA δ Lib ∞	VARIABLE 2.3 DAYS 4.8-5.9
	R Lib	VARIABLE 242 DAYS 9.8-15.9
	S Lib	VARIABLE 193 DAYS 7.5-13.0
	U Lib	VARIABLE 226 DAYS 9.0-15.0
	Y Lib	VARIABLE 275 DAYS 7.6-14.7
	RS Lib	VARIABLE 217 DAYS 7.0-13.0
	MU μ Lib ⊕	DOUBLE 5.8 & 6.7
	IOTA I Lib ⊕	DOUBLE 4.7 & 9.7

NORTH STAR

+90° +80° +70° +60° +50° +40° +30° +20° +10° 0° SKY EQUATOR -10° -20° -30° -40° -50°

-18°

HORIZON LINE

NORTH   LIBRA   SOUTH

EQUATOR   EARTH AXIS

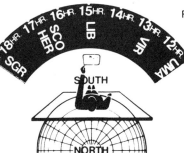

For DAYLIGHT SAVING TIMES see note on page 10.

1 MAY MIDNIGHT
15 MAY 11 P.M.
1 JUN. 10 P.M.
15 JUN. 9 P.M.
1 JUL. 8 P.M.

A one-hour bus or train ride away from town or city can take you to the grandest free show on earth. Pack a picnic, take blankets and don't forget a compass, this booklet and a red-covered flashlight. You are in for a rare treat. Your first observing trip to a dark-sky area will not be your last.

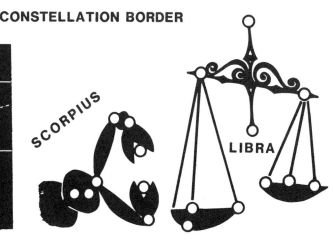

Libra is one of the Zodiac constellations. The sun passes through it from November 1 to November 22.

To the ancient Romans, a *libra* was a unit of weight. Even the Greeks did not regard the stars in this field as a balance. To them they were part of nearby Scorpius. The name of Libra's brightest star, "Zuben Elgenubi," confirms the Scorpion connection. Translated from the Arabic, it meant "the right claw" (of the Scorpion). "Zuben Elschamali" referred to the left claw. "Zuben Elgenubi" is sometimes abbreviated into "Zuben'ubi."

Later, with Julius Caesar and his "Julian Calendar," Libra was added as the twelfth constellation of the Zodiac, and Scorpius was reduced to its present size. At first the borders of constellations were not defined very clearly, and the lines separating them were vaguely drawn at random. This can be seen in some of the Bode atlas illustrations, where faint dotted lines separate one starfield from another. Only in this century have astronomers from all over the world established accurate borders by international treaty.

Purchase of a good sky atlas is a worthwhile investment for serious astronomers. As can be seen below, star maps today accurately mark the borders of constellations, and identify the brighter stars with Greek letters and the fainter ones with numbers, showing their relative magnitudes with different-sized dots. An atlas will show all celestial objects in their accurate positions. Alpha (α) Librae, for example, can be pinpointed by referring to R.A. and Dec. lines for precise coordinates. No Messier discoveries appear in Libra, or they would be shown along with many fainter objects.

Also note the positions of the variables R, S, T, and RS Librae.

DEGREES IN DECLINATION

HOURS IN RIGHT ASCENSION

POSITION OF ALPHA LIBRAE: **14 HR 50M −16°**

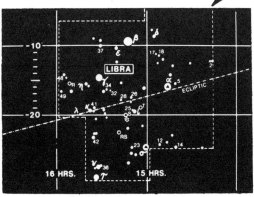

STARATLAS DETAILS

CONSTELLATION BORDER

SCORPIUS

LIBRA

63

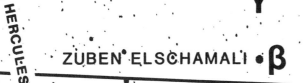

HERCULES PAGE 72

VIRGO PAGE 60

−10°

ZUBEN ELSCHAMALI •β

•δ

Υ

R

•γ

•μ

η

ECLIPTIC

θ

ZUBEN ELGENUBI α •

PATH OF THE SUN IN NOVEMBER

I

U

S

SCORPIUS PAGE 68

RS

LIBRA
α Lib 14h51m −16°02'

•σ

u

15h

64

Find the pictogram stars on this page and connect them with ruled
colored lines. You can also create your own ideograms with red markings
which will filter out and disappear under a red flashlight (page 135).

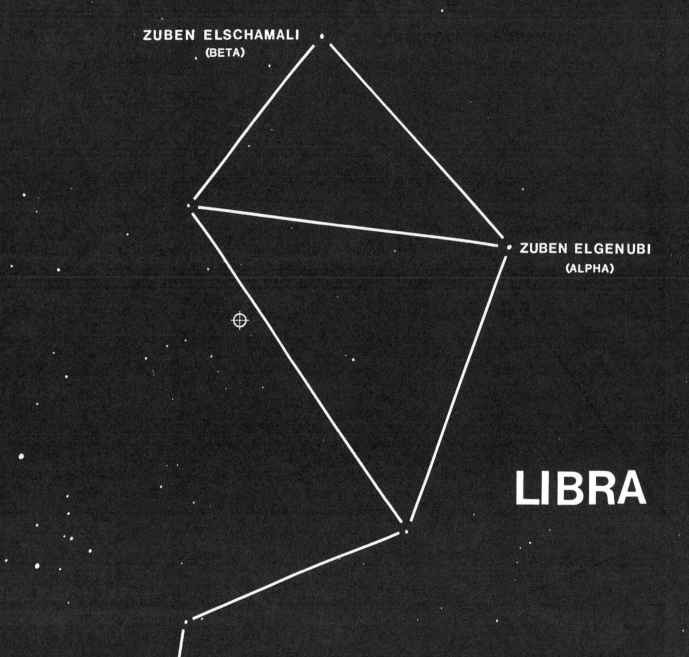

ALIGN WITH ZUBEN ELGENUBI
AND ZUBEN ELSCHAMALI

N

STANDARD 50 MM LENS

ZUBEN ELSCHAMALI
(BETA)

ZUBEN ELGENUBI
(ALPHA)

LIBRA

ANTARES
(ALPHA SCORPII)

AIM CAMERA AT ⊕ TARGET

65

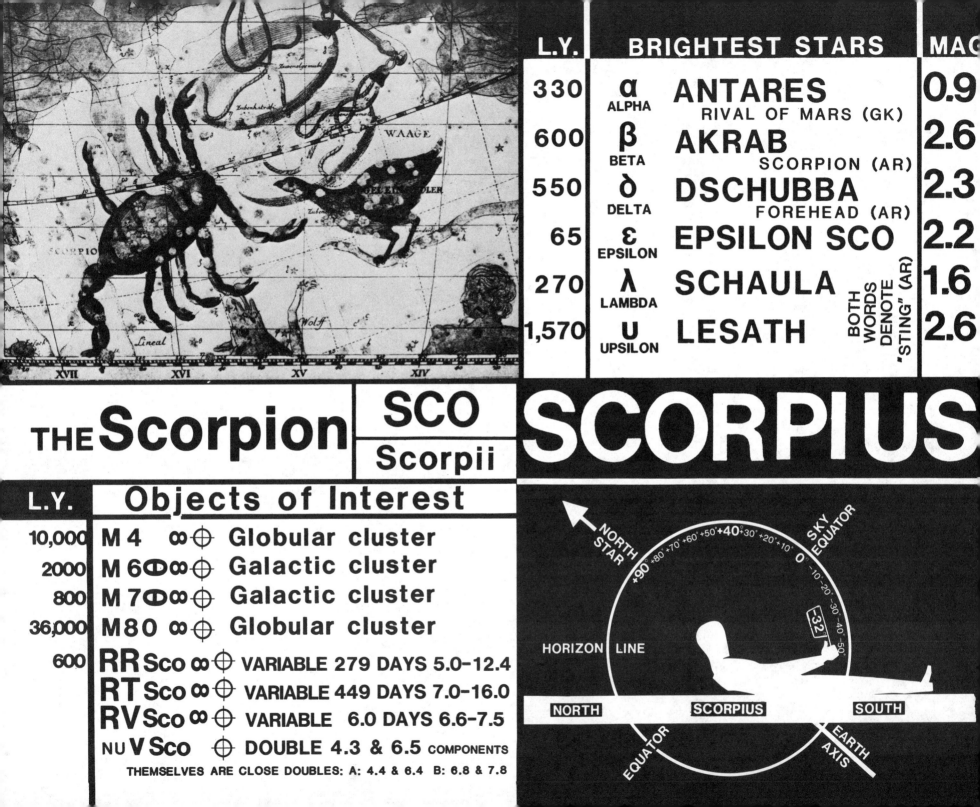

L.Y.	BRIGHTEST STARS		MAG
330	α ALPHA	ANTARES RIVAL OF MARS (GK)	0.9
600	β BETA	AKRAB SCORPION (AR)	2.6
550	δ DELTA	DSCHUBBA FOREHEAD (AR)	2.3
65	ε EPSILON	EPSILON SCO	2.2
270	λ LAMBDA	SCHAULA	1.6
1,570	U UPSILON	LESATH	2.6

BOTH WORDS DENOTE "STING" (AR)

THE Scorpion

SCO
Scorpii

SCORPIUS

L.Y.	Objects of Interest		
10,000	M 4 ∞⊕	Globular cluster	
2000	M 6⊕∞⊕	Galactic cluster	
800	M 7⊕∞⊕	Galactic cluster	
36,000	M 80 ∞⊕	Globular cluster	
600	RR Sco ∞⊕	VARIABLE 279 DAYS 5.0-12.4	
	RT Sco ∞⊕	VARIABLE 449 DAYS 7.0-16.0	
	RV Sco ∞⊕	VARIABLE 6.0 DAYS 6.6-7.5	
	NU V Sco ⊕	DOUBLE 4.3 & 6.5 COMPONENTS	

THEMSELVES ARE CLOSE DOUBLES: A: 4.4 & 6.4 B: 6.8 & 7.8

NORTH STAR

+90° +80° +70° +60° +50° +40° +30° +20° +10° 0° -10° -20° -30° -40° -50°

SKY EQUATOR

132°

HORIZON LINE

NORTH   SCORPIUS   SOUTH

EQUATOR   EARTH AXIS

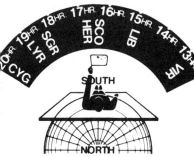

For DAYLIGHT SAVING TIMES see note on page 10.

1 JUN. MIDNIGHT
15 JUN. 11 P.M.
1 JUL. 10 P.M.
15 JUL. 9 P.M.
1 AUG. 8 P.M.

Plan a night or two in the country with friends or loved ones. Check the newspaper for a weekend nearest the new moon, when it will be dark all night long. Plan for a campout under the stars. You will see a grand show. Admission is free.

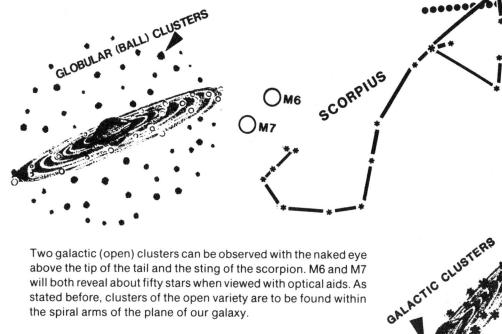

GLOBULAR (BALL) CLUSTERS

M6

M7

SCORPIUS

Scorpius dates back to when the Zodiac consisted of only six constellations. The sky hieroglyphics at the Egyptian Temple of Denderah show a scorpion as it appears on more recent charts. In ancient China the constellation was known as the "Azure Dragon."

The sun is in Scorpius for only seven days, from November 23 to the 30th. Then, before moving into Sagittarius, the sun will travel through the constellation of Ophiuchus, the Serpent Bearer, from November 30 until December 18. When astrology was invented some two thousand years ago, the sun did not pass through Ophiuchus (illustrated in the Bode atlas print on page 20). Today, however, due to a very slow earth-wobble called precession (page 99), the sun spends more than twice as much time with the Snake than with the Scorpion.

The brightest star in Scorpius is Antares. It was often called "Rival of Mars" because of its reddish color, which is reminiscent of our neighboring planet. Alpha (α) Scorpii is a double star, the companion having a magnitude of 6. Because the Alpha (α) star is so bright, an 8-inch telescope and good seeing are needed to split the pair.

If you observe from an area where skies are good and dark, the Milky Way will present an extraordinary panorama, leading the eye to the left and north toward the "teapot" of Sagittarius. This is one of the most beautiful regions in the vault of the heavens.

Two galactic (open) clusters can be observed with the naked eye above the tip of the tail and the sting of the scorpion. M6 and M7 will both reveal about fifty stars when viewed with optical aids. As stated before, clusters of the open variety are to be found within the spiral arms of the plane of our galaxy.

Two fine examples of another type of cluster can be observed close to Antares. M4 lies just to the right of the red giant and might be spotted with the naked eye, appearing as a 6th-magnitude star. M80, farther north, is a fainter object and much more compact. Both are best viewed with telescopes to reveal the beauty of these typical "globular clusters." These ball-shaped groupings of stars lie above and below the saucer-flat disc of our galaxy. We know about a hundred such "globulars," each containing tens or hundreds of thousands of stars.

GALACTIC CLUSTERS

OUR GALAXY

GLOBULAR CLUSTER

M 4

GALACTIC CLUSTER

M 6

GALACTIC CLUSTER

M 7

GLOBULAR CLUSTER

M 80

OPHIUCHUS

PATH OF THE SUN IN DECEMBER

HERCULES PAGE 72

ECLIPTIC

ν β
ω
δ

M80

σ
ANTARES α
π

T
M4

LIBRA PAGE 64

ρ

SCORPIUS

αSco 16h29m -26°26'

-30°

RR

M6

M7

RV
ε

RT

M4

SCHAULA λ U

SAGITTARIUS PAGE 76

K

ι

θ

η

ζ

μ

Find the pictogram stars on this page and connect them with ruled colored lines. You can also create your own ideograms with red markings which will filter out and disappear under a red flashlight (page 135).

16h

ALIGN WITH ANTARES AND SCHAULA

N

SCORPIUS

ANTARES
(ALPHA)

SCHAULA
(L'AMBDA)

STANDARD 50 MM LENS

69

KNEELING Man

HERCULES

BRIGHTEST STARS

L.Y.		BRIGHTEST STARS	MAG
220	α ALPHA	RAS ALGHETI HEAD OF THE KNEELER (AR)	3.2
105	β BETA	KORNEPHOROS CLUB BEARER (GK)	2.8
140	γ	GAMMA HER	3.8
91	δ	DELTA HER	3.1
31	ζ	ZETA HER	2.8
230	λ LAMBDA	MAASYM WRIST (AR)	4.4

Objects of Interest

L.Y.			
25 000	M 13 ∞ ⊕	Great Hercules cluster	
30,000	M 92 ∞ ⊕	Globular cluster	
	S Her ∞ ⊕	VARIABLE 307 DAYS 6.4-13.8	
	T Her ∞ ⊕	VARIABLE 165 DAYS 6.8-13.9	
	U Her ∞ ⊕	VARIABLE 406 DAYS 6.5-13.4	
220	ALPHA α Her ⊕	IRREGULAR VARIABLE 3.0-4.0 AND DOUBLE STAR 3-4 & 5.4	
	KAPPA K Her ⊕	DOUBLE 5.0 & 6.0	

NORTH STAR

SKY EQUATOR

+90° +80° +70° +60° +50° +40° +30° +20° +10° 0° -10° -20° -30° -40° -50°

+31°

HORIZON LINE

NORTH HERCULES SOUTH

EQUATOR

EARTH AXIS

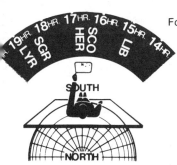

For DAYLIGHT SAVING TIMES see note on page 10.

1 JUN. MIDNIGHT
15 JUN. 11 P.M.
1 JUL. 10 P.M.
15 JUL. 9 P.M.
1 AUG. 8 P.M.

At this time of the year starlovers meet at star parties, under favorable skies away from light-pollution. Such get-togethers are usually held on new-moon weekends, when the moon rises and sets with the sun and when the nights are dark. Phone the Earth Science department of your nearest college. They will know where and when stargazers meet.

HERCULES

M 13

Heracles, hero and demigod to the ancient Greeks, kneels immortalized in the summer sky. The legendary Roman name of Hercules was bestowed on a constellation already known in Babylonian days as the "Kneeling One."

The pictogram does not allow us to recognize the figure as it is portrayed by Bode. In the culminating position the "one who kneels" does so in an upside-down position. It helps little to remember that twelve hours, or six months, from now the figure will be right side up, because by then Hercules will be in the daytime sky and invisible altogether.

The central "Butterfly" in the constellation and the northern part of it, called the "keystone" of Hercules, will help us find the starfield and to relate it to the bright star, Alpha (α) Herculis, which is said to mark the head of the kneeling figure. Ras Algethi is a many-splendored star. What appears as a red giant of about 3rd magnitude is in fact a pair of stars. The components are of magnitude 3.0 and 5.4 respectively. The brighter 3rd-magnitude star is a variable, ranging from magnitude 3.0 to 4.0 without following any regular schedule, as many stars do. Thus Alpha (α) Her can be described as an irregular variable telescopic double star. Do not expect to be able to notice the very slight changes in magnitude until you have had ample practice with stars of more impressive variations (page 121).

The "Keystone" pictogram is made up of the stars Epsilon (ε), Zeta (ζ), Eta (η), and Pi (π) Herculis. It can help us find the most famous component of the constellation, the Great Globular Cluster in Hercules. On the line connecting Eta (η) Her and Zeta (ζ) Her, about one third of the way from Eta (η), lies the spectacular sphere of countless suns. For a while it was called "Halley's Nebula," after its reputed discoverer, of comet fame. Messier entered it as number 13, "a nebula without stars." Today we know M13 as the most spectacular of all star clusters in the northern skies. Even if we will never know just how many stars it contains, its diameter overall is known to be about 200 light-years. The core, where the light of a million suns blends into a blaze of brilliance, is 100 light-years wide. Multiplying width by height and by depth (100 LY x 100 LY x 100 LY), we obtain a supercube with a volume of a million cubic light-years. With one star for each of these light-year cubes, suns would still be few and far between.

If M13 had the diameter of our earth, then each of its stars would be about the size of a marble. The distance between the tiny marbles might be 30 miles at the center, up to eighty miles nearer to the perimeter. Even where we see an unfathomable density of teeming suns, eighty miles is a long way between marbles.

MARBLE
ACTUAL SIZE

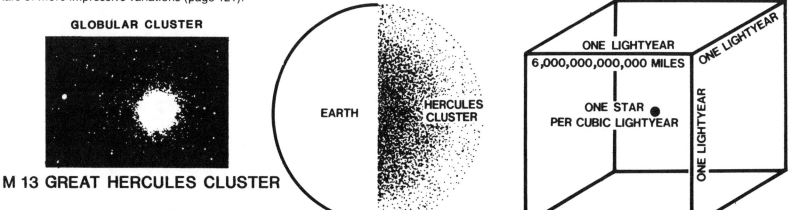

GLOBULAR CLUSTER

M 13 GREAT HERCULES CLUSTER

EARTH

HERCULES CLUSTER

ONE LIGHTYEAR
6,000,000,000,000 MILES
ONE LIGHTYEAR

ONE STAR PER CUBIC LIGHTYEAR

ONE LIGHTYEAR

Find the pictogram stars on this page and connect them with ruled colored lines. You can also create your own ideograms with red markings which will filter out and disappear under a red flashlight (page 135).

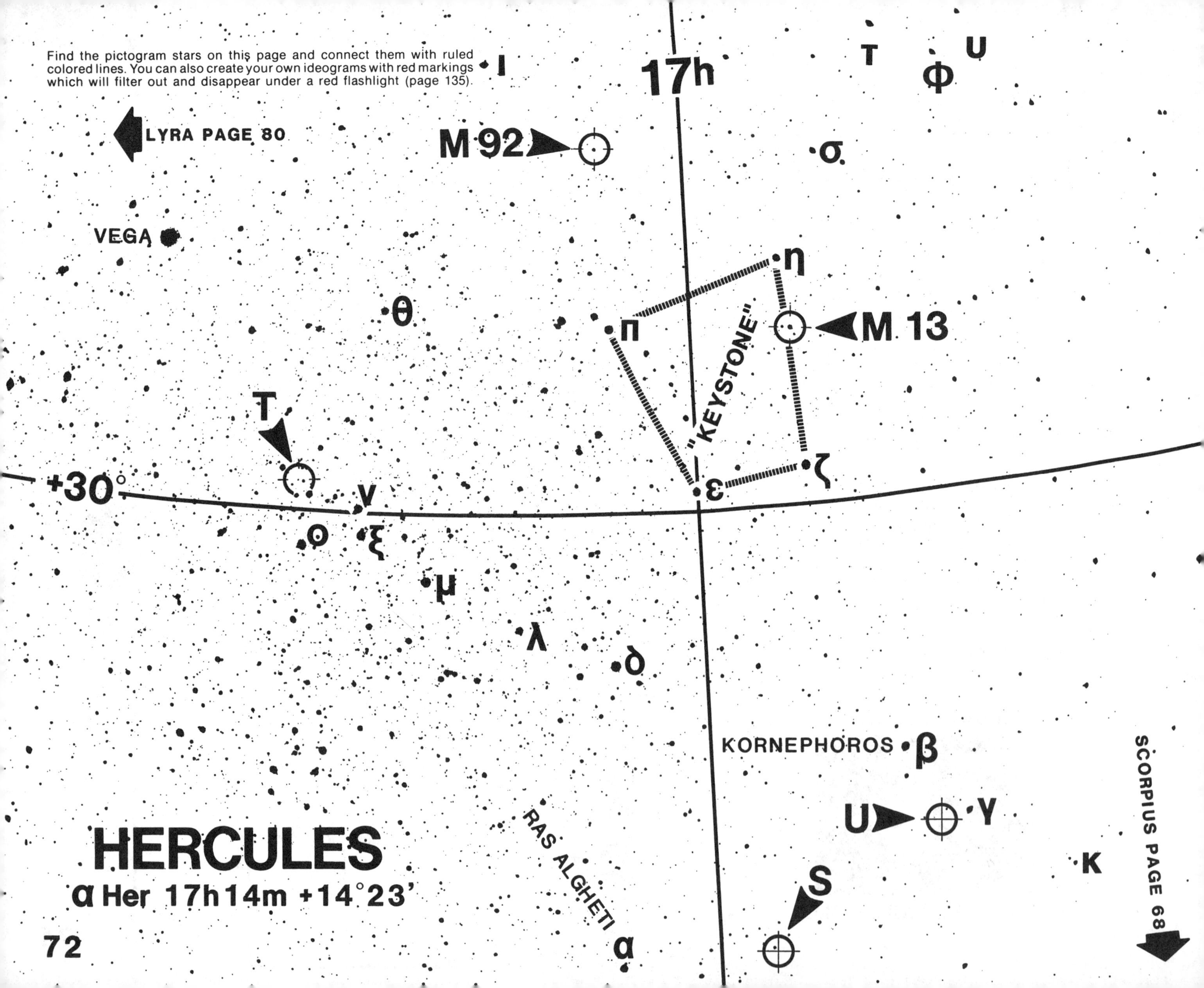

◀ LYRA PAGE 80

I

17h

T Φ U

M 92 ▶ ⊕

σ

VEGA ●

θ

η

Π

"KEYSTONE" ⊕ ◀ M 13

T ▼

ζ

+30° ε

ν

ο ξ

μ

λ

δ

KORNEPHOROS ● β

RAS ALGHETI

U ▶ ⊕ ● γ

HERCULES

α Her 17h 14m +14°23'

S ▼

⊕

K

SCORPIUS PAGE 68 ▼

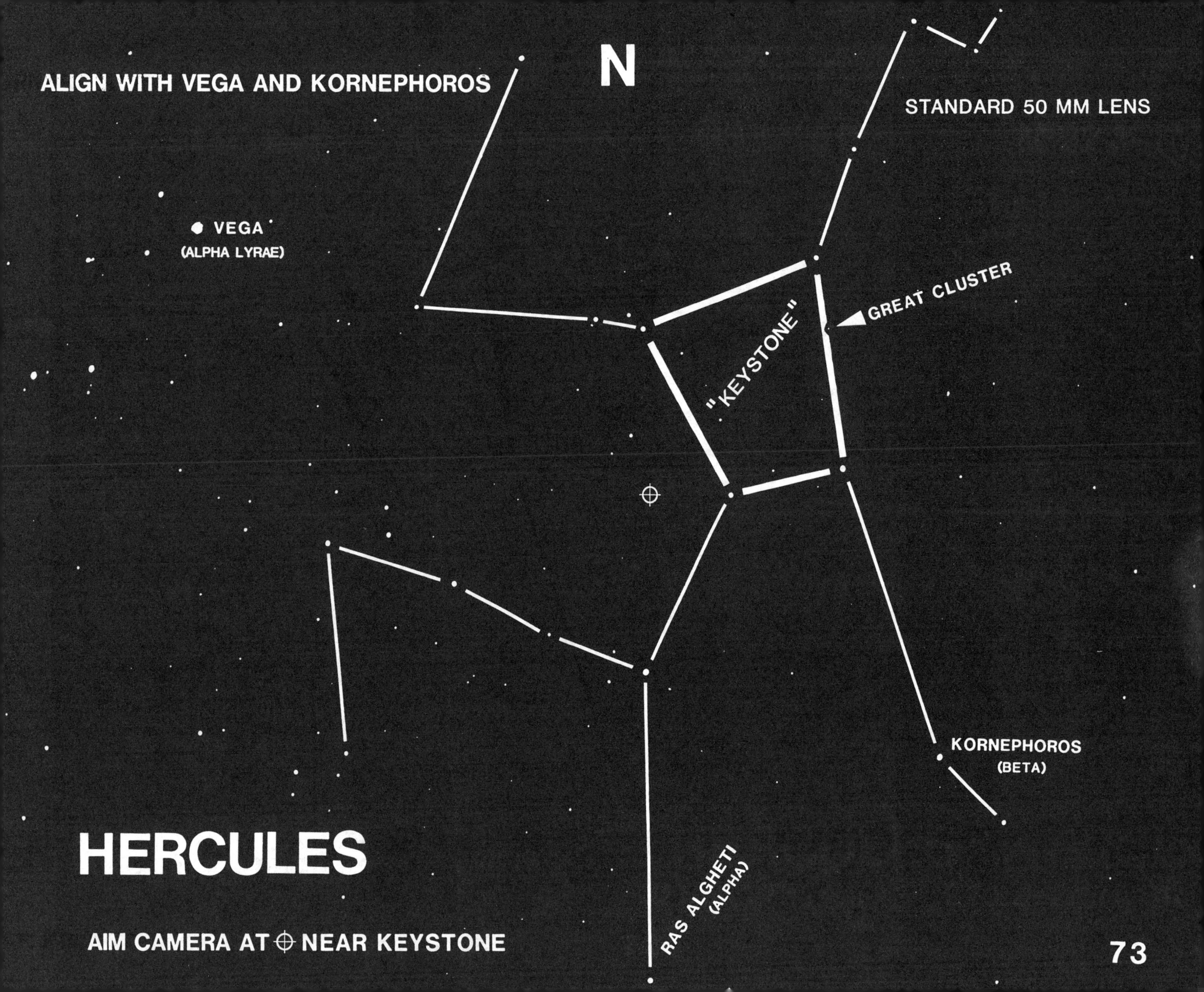

ALIGN WITH VEGA AND KORNEPHOROS

N

STANDARD 50 MM LENS

● VEGA
(ALPHA LYRAE)

"KEYSTONE"

GREAT CLUSTER

KORNEPHOROS
(BETA)

HERCULES

RAS ALGHETI
(ALPHA)

AIM CAMERA AT ⊕ NEAR KEYSTONE

73

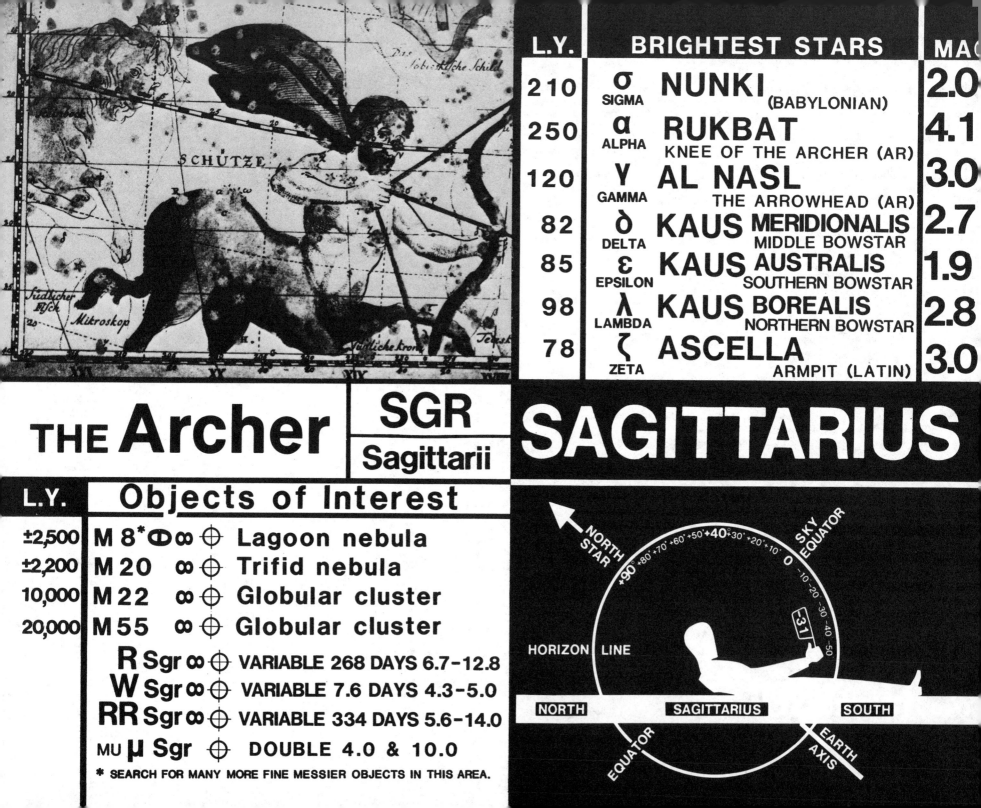

L.Y.	BRIGHTEST STARS		MAG
210	σ SIGMA	NUNKI (BABYLONIAN)	2.0
250	α ALPHA	RUKBAT KNEE OF THE ARCHER (AR)	4.1
120	γ GAMMA	AL NASL THE ARROWHEAD (AR)	3.0
82	δ DELTA	KAUS MERIDIONALIS MIDDLE BOWSTAR	2.7
85	ε EPSILON	KAUS AUSTRALIS SOUTHERN BOWSTAR	1.9
98	λ LAMBDA	KAUS BOREALIS NORTHERN BOWSTAR	2.8
78	ζ ZETA	ASCELLA ARMPIT (LATIN)	3.0

THE Archer

SGR Sagittarii

SAGITTARIUS

L.Y.	Objects of Interest		
±2,500	M 8* ⚭ ∞ ⊕	Lagoon nebula
±2,200	M 20 ∞ ⊕	Trifid nebula
10,000	M 22 ∞ ⊕	Globular cluster
20,000	M 55 ∞ ⊕	Globular cluster
	R Sgr ∞ ⊕	VARIABLE 268 DAYS 6.7-12.8
	W Sgr ∞ ⊕	VARIABLE 7.6 DAYS 4.3-5.0
	RR Sgr ∞ ⊕	VARIABLE 334 DAYS 5.6-14.0
	MU μ Sgr ⊕	DOUBLE 4.0 & 10.0

* SEARCH FOR MANY MORE FINE MESSIER OBJECTS IN THIS AREA.

NORTH STAR

+90° +80° +70° +60° +50° +40° +30° +20° +10° 0 SKY EQUATOR -10 -20 -30 -40 -50

-31

HORIZON LINE

NORTH — SAGITTARIUS — SOUTH

EQUATOR

EARTH AXIS

For DAYLIGHT SAVING TIMES see note on page 10.

1 JUL. MIDNIGHT
15 JUL. 11 P.M.
1 AUG. 10 P.M.
15 AUG. 9 P.M.
1 SEP. 8 P.M.

Look for Perseid meteors from August 1 to 15. Don't miss the night of August 12. Plan for a trip away from town or city. Take blankets or sleeping bags. Open a camera to the night sky. You may catch a shooting star or even a fireball. It should be a night to remember.

SAGITTARIUS

Studies of the positions of globular clusters, in relation both to us and to our galaxy, have enabled astronomers to establish that our solar system lies rather far out nearer the perimeter of our galaxy. We know that the "galactic center," as seen from here, lies in the direction of Sagittarius. That is why the Milky Way appears most dazzling in the region of the Archer, where it glows with uncountable stars. A trip away from city lights will be a rewarding delight some moonless summer night. You will see nature's most spectacular show, with perhaps the accompaniment of chirping crickets or softly hooting owls.

Sagittarius is a southern Zodiac constellation, and the sun passes through it from December 19 to January 19. Even city dwellers should be able to see the "Teapot" pictogram. Nunki, Sigma (σ) Sagitarii, forms the "top of the handle," and Al Nasl, Gamma (γ) Sgr, is the "front of the spout." Al Nasl is also the tip of the arrow of the archer which makes the bow recognizable. Heaven only knows where the ancients found the mythological centaur, half man and half horse, as pictured in Bode's atlas. But then, have you ever seen a centaur?

Even with an inferior telescope Messier catalogued no fewer than fifteen interesting objects in this region. You may want to record the surrounding areas with your camera and later survey your own discoveries with a telescope.

M8, the famous Lagoon nebula, just above the spout of the teapot, is one of the brightest treasures in Sagittarius. It reveals itself in color photographs with its telltale signature of pink hydrogen gas. Just like the Orion nebula, this swirling region is a birthplace of new stars.

M20, the Trifid nebula, lies less than two degrees to the north of M8. It is the fainter of the two objects and derives its name from dark dust lanes which seem to divide glowing clouds of hydrogen gas into three parts.

M22 and M55 are globular clusters. They are of similar apparent magnitude and will appear as 6th-magnitude stars to the naked eye or to the wide-angle camera.

W Sgr and RR Sgr are the two brightest variable stars in Sagittarius, with others well within your reach. There are about 2,300 variables in this constellation alone. Many need sensitive photometers and powerful telescopes to record their fractional changes in magnitude.

BIRTHPLACE OF NEW STARS

M 8 LAGOON NEBULA

GLOWING NEBULA

M 20 TRIFID NEBULA

GLOBULAR CLUSTER

M 22

GLOBULAR CLUSTER

M 55

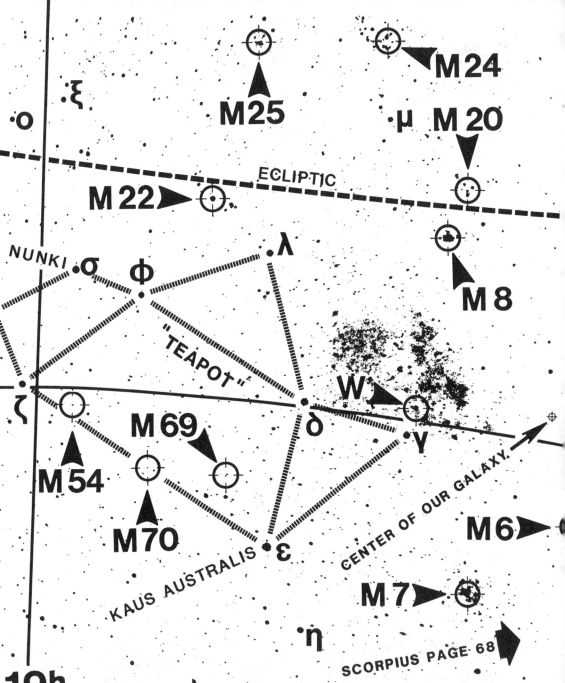

Find the pictogram stars on this page and connect them with ruled colored lines. You can also create your own ideograms with red markings which will filter out and disappear under a red flashlight (page 135).

R

π
ο
ξ

M25

M24

μ M20

PATH OF THE SUN IN JANUARY

ECLIPTIC

M22

SAGITTARIUS

σ Sgr 18h55m −26°18'

NUNKI σ φ λ

M8

τ

"TEAPOT"

RR

ζ

W

γ

−30

M55

M69

δ

M54

M70

ε

CENTER OF OUR GALAXY

M6

KAUS AUSTRALIS

θ

M7

η

CAPRICORN PAGE 88

SCORPIUS PAGE 68

19h

α

RU

76

EXPLORE THE MANY MESSIER OBJECTS IN THIS RICH REGION. THE CLUSTERS M6 AND M7 BELONG TO ADJACENT SCORPIUS.

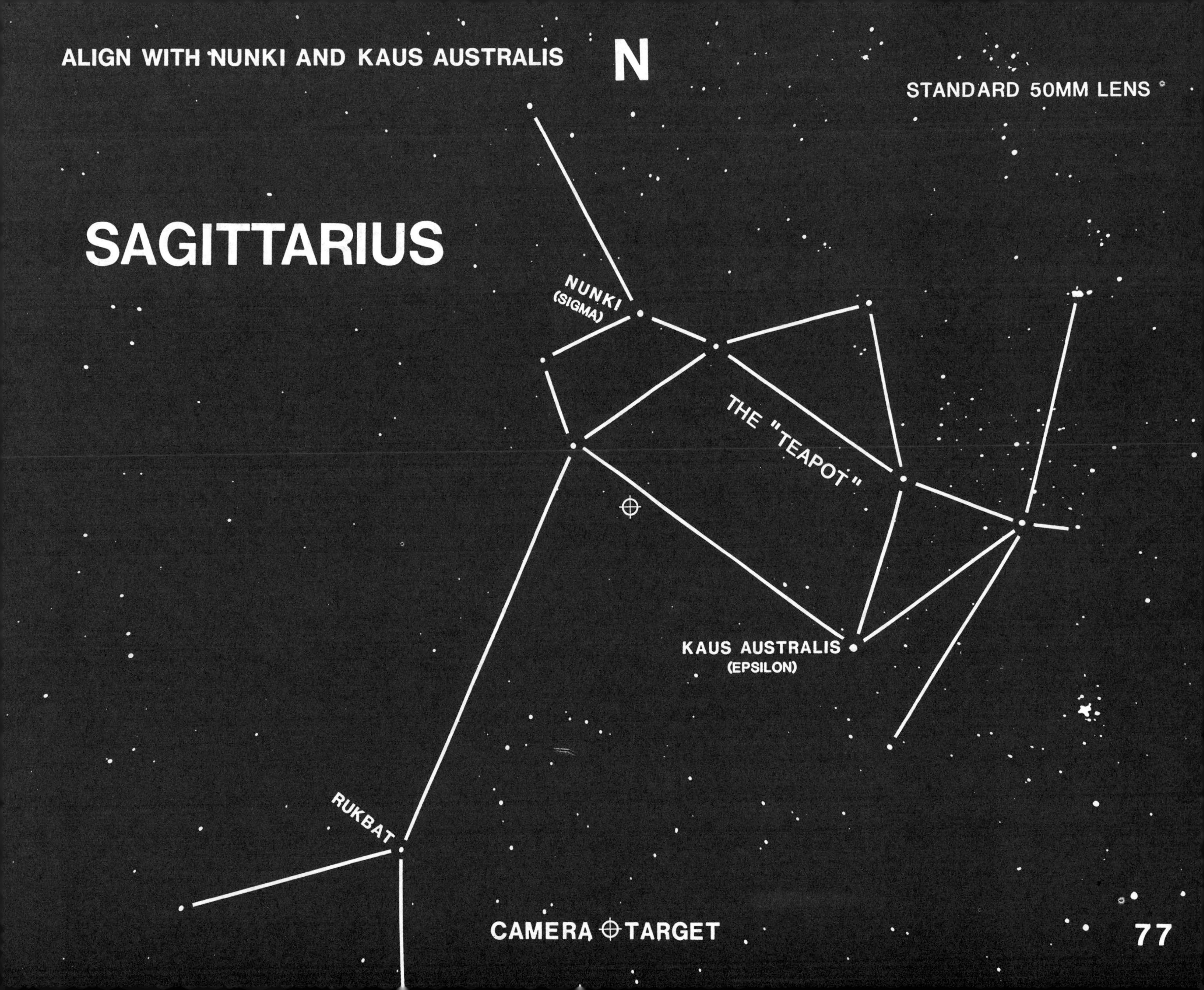

ALIGN WITH NUNKI AND KAUS AUSTRALIS

N

STANDARD 50MM LENS

SAGITTARIUS

NUNKI
(SIGMA)

THE "TEAPOT"

KAUS AUSTRALIS
(EPSILON)

RUKBAT

CAMERA ⊕ TARGET

THE Lyre

LYR
Lyrae

LYRA

L.Y.	BRIGHTEST STARS		MAG
26	α VEGA ALPHA	NASR AL-WAQI SWOOPING EAGLE (AR)	0.0
300	β SHELIAK BETA	THE HARP (AR)	3.5
190	γ SULAPHAT GAMMA	THE TURTLE (AR)	3.2
180	ε EPSILON LYR		SEE BELOW
130	ζ ZETA LYR		4.0
800	η ALADFAR ETA	THE TALONS (AR)	4.4

Objects of Interest

L.Y.			
40,000	M56 ⊕	Globular cluster	
±2,000	M57 ⊕	Ring nebula (PLANETARY)	
	R Lyr ∞ ⊕	VARIABLE 46 DAYS 3.8–5.0	
	W Lyr ∞ ⊕	VARIABLE 196 DAYS 7.3–13.0	
	BETA β Lyr ∞ ⊕	VARIABLE 13 DAYS 3.3–4.3	
	EPSILON ε Lyr ∞ ⊕	DBL. DOUBLE 4.7 & 4.5	
	ε¹ ⊕ 5.1 & 6.0	ε² ⊕ 5.1 & 5.4	
	ETA η Lyr ⊕	DOUBLE 4.5 & 8.7	

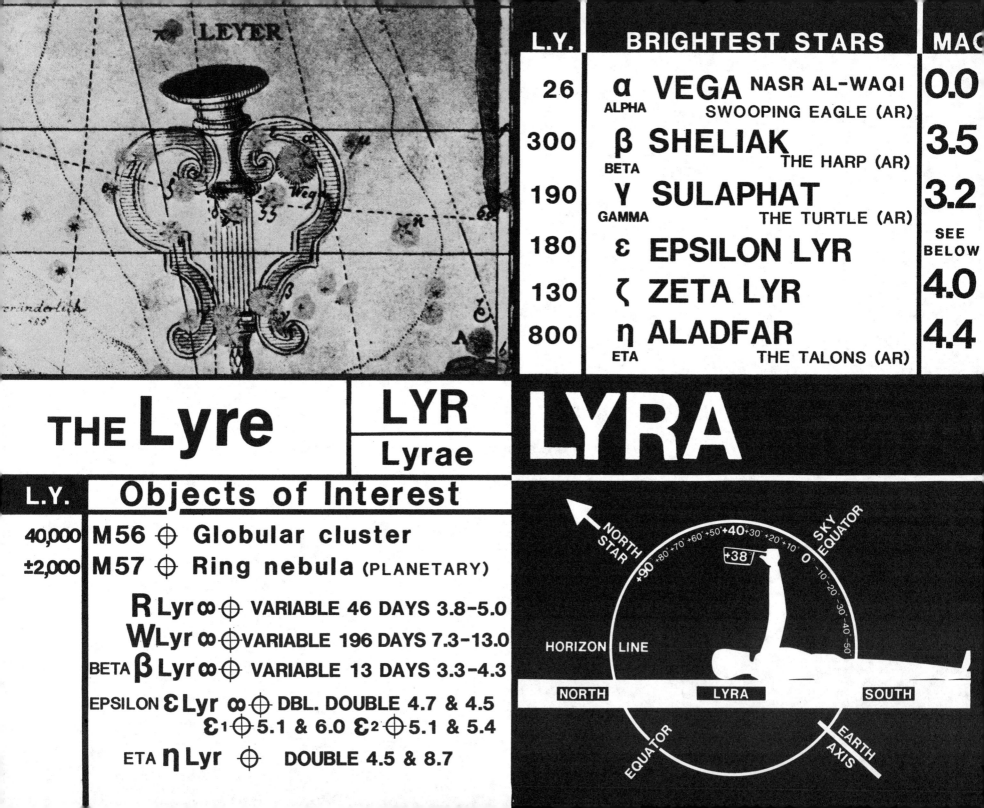

NORTH STAR

+90 +80 +70 +60 +50 +40 +30 +20 +10 0 SKY EQUATOR

+38

−10 −20 −30 −40 −50

HORIZON LINE

NORTH   LYRA   SOUTH

EQUATOR   EARTH AXIS

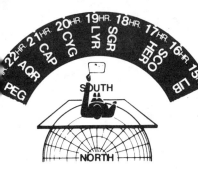

For DAYLIGHT SAVING TIMES see note on page 10.

1 JUL. MIDNIGHT
15 JUL. 11 P.M.
1 AUG. 10 P.M.
15 AUG. 9 P.M.
1 SEP. 8 P.M.

Have you thought about using the coming winter to build your own telescope? Subscribe to astronomy magazines now (see page 142, "Resources") and pick up books at your library to show you how easy it can be to build your own optical "window to the sky."

The constellation of Lyra and the Sagittarius starfield occupy approximately the same melon slice of right ascension. While the Archer lies below the celestial equator, Lyra soars at +40°. As illustrated on page 12, this +40° position is directly overhead for observers near the latitude of +40° on earth. At the indicated time of culmination, the fifth-brightest star in the entire celestial sphere will be easy to identify in the "zenith" position. You will be looking at Vega, named the "Swooping Eagle," also called "Harp Star" after the constellation itself.

A lyre is a musical instrument with strings, not unlike the harp. It was often played in ancient Greece. The celestial Lyra is associated with Hermes, Orpheus and Apollo, with gods and heroes. Its music charmed all who heard it.

Alpha (α) Lyrae is said to be the first star ever to be photographed at Harvard Observatory, in July 1850. The exposure time was one hundred seconds. Today anyone with a camera can do much better (pages 127-29). Try to capture its light on color film with a time exposure of just a few seconds. First try a four-second exposure. Next, double this and shoot for eight seconds. Redouble and expose for sixteen seconds. Finally, try thirty seconds and then one minute. Welcome to the ranks of astrophotographers! You will soon find the fainter stars starting to appear in the longer exposures of your slides or prints and will realize that

however sturdy your tripod, it was standing on a turning planet. Inevitably, you will compare your own photos with the ones overleaf and realize that there really is no big mystery to star photography. Mark "North" on your photos to establish a standard format for your own collection of starlight.

If you have a telephoto lens, Epsilon (ε) Lyrae will reveal itself as a double star, with two components of magnitudes 4.5 and 6.5, called Epsilon 1 and Epsilon 2. With a telescope both of these can be split into yet two more stars. Epsilon (ε) Lyr is a "double double" star.

Do not expect, at first, to be able to record M57, the famous Ring nebula. It is a faint 9th-magnitude object glowing hydrogen pink. It needs longer exposures and longer lenses. Such expanding spheres of gas are the remnants of a central star which blew off its outer layers. To some early astronomers such round objects may have suggested faint planets, which is why they are called "planetary nebulae." They have no connection with planets, however.

We can see only the "walls" of such remaining bubbles of gas, where they are deepest and thickest. With a fair telescope the Ring nebula can readily be found by systematically scanning the area shown in the starfield.

LYRA α
ε EPSILON ALPHA
VEGA
δ DELTA
M 57
M56

M 57

GLOBULAR CLUSTER

M 56

MESSIER LOG-ENTRY

January 23, 1779
V.56 Nebula without Stars. Discovered on same day as the Comet of 1779 on Jan. 19 Lyra. Position: 18.0 ~ +29.48.14

PLANETARY NEBULA

M 57 RING NEBULA

TO EARTH

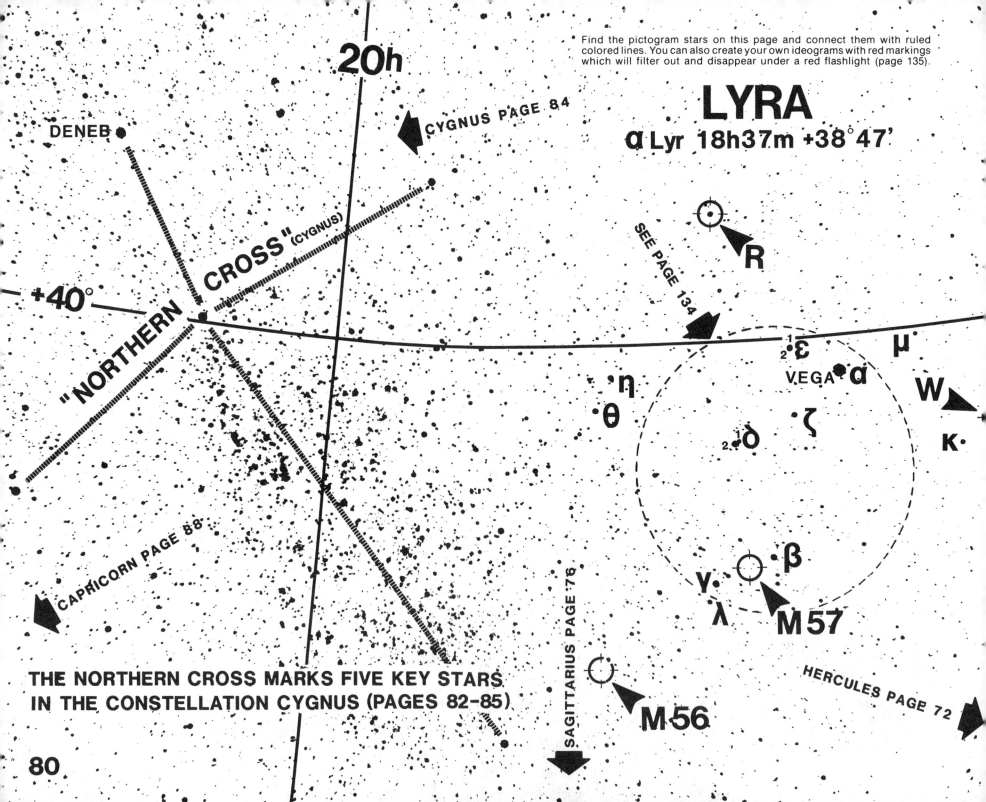

Find the pictogram stars on this page and connect them with ruled colored lines. You can also create your own ideograms with red markings which will filter out and disappear under a red flashlight (page 135).

20h

CYGNUS PAGE 84

LYRA
α Lyr 18h37m +38°47'

DENEB

"NORTHERN CROSS" (CYGNUS)

+40°

SEE PAGE 134

R

VEGA α

μ

W

η ι

ζ

θ

δ

K

CAPRICORN PAGE 88

γ

β

λ

M 57

SAGITTARIUS PAGE 76

HERCULES PAGE 72

M 56

THE NORTHERN CROSS MARKS FIVE KEY STARS IN THE CONSTELLATION CYGNUS (PAGES 82-85)

80

LYRA

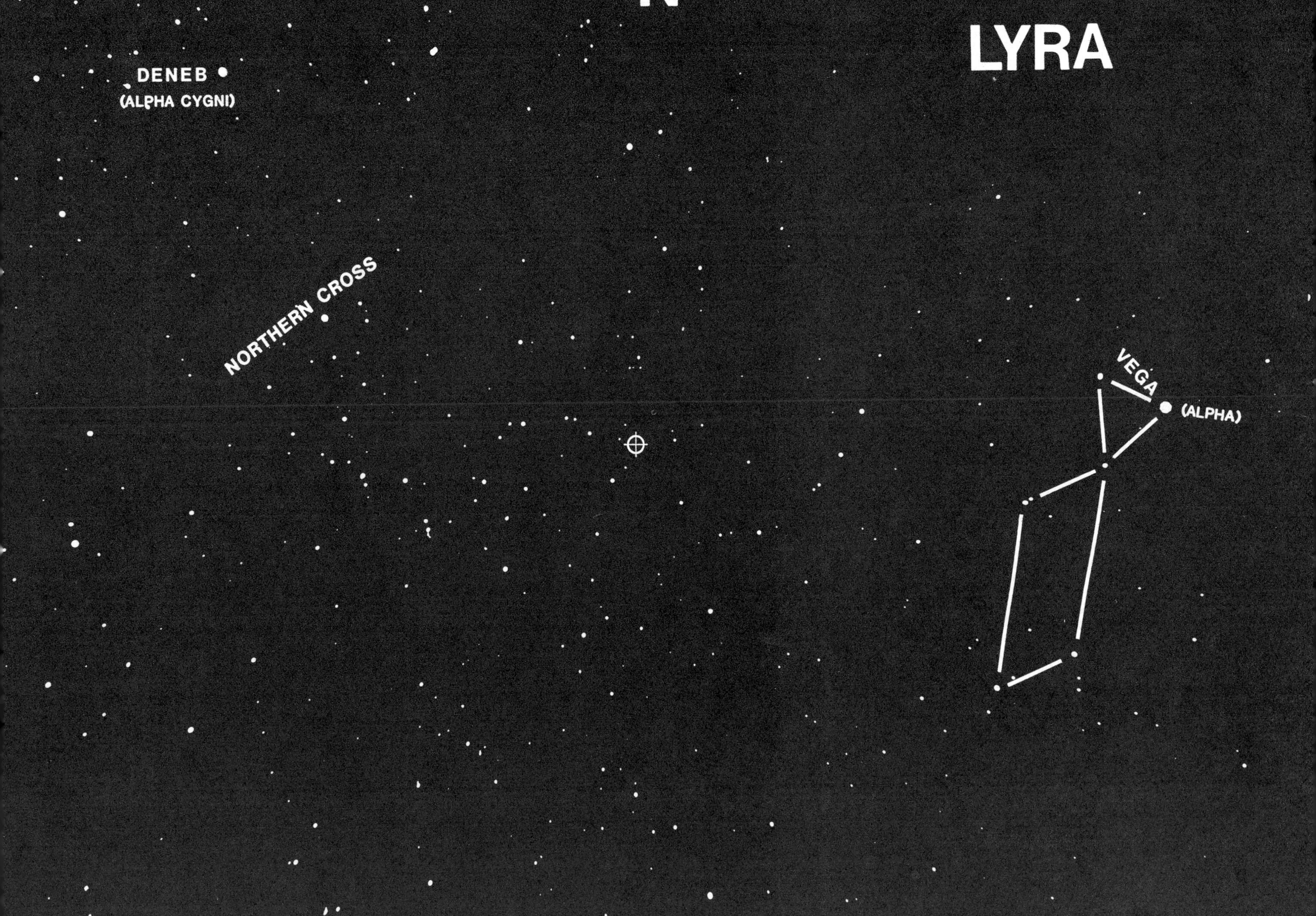

DENEB •
(ALPHA CYGNI)

NORTHERN CROSS

VEGA

(ALPHA)

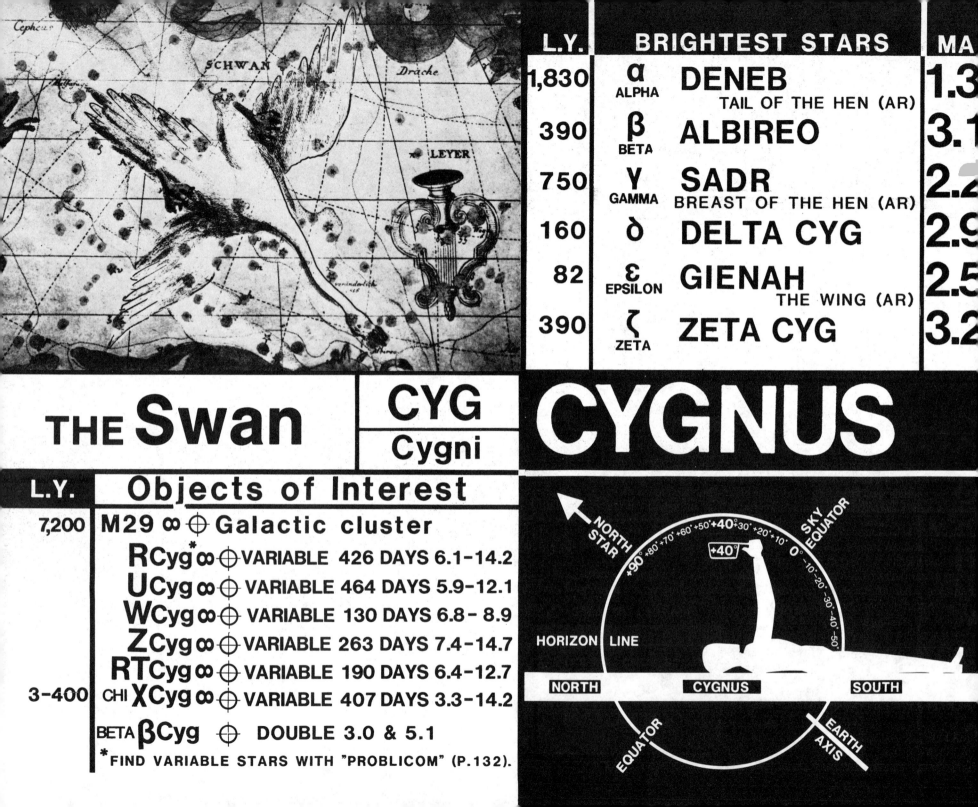

L.Y.	BRIGHTEST STARS		MA
1,830	α ALPHA	DENEB TAIL OF THE HEN (AR)	1.3
390	β BETA	ALBIREO	3.1
750	γ GAMMA	SADR BREAST OF THE HEN (AR)	2.2
160	δ	DELTA CYG	2.9
82	ε EPSILON	GIENAH THE WING (AR)	2.5
390	ζ ZETA	ZETA CYG	3.2

THE Swan

CYG
Cygni

CYGNUS

L.Y.	Objects of Interest
7,200	M29 ∞ ⊕ Galactic cluster
	RCyg* ∞ ⊕ VARIABLE 426 DAYS 6.1-14.2
	UCyg ∞ ⊕ VARIABLE 464 DAYS 5.9-12.1
	WCyg ∞ ⊕ VARIABLE 130 DAYS 6.8-8.9
	ZCyg ∞ ⊕ VARIABLE 263 DAYS 7.4-14.7
	RTCyg ∞ ⊕ VARIABLE 190 DAYS 6.4-12.7
3-400	CHI XCyg ∞ ⊕ VARIABLE 407 DAYS 3.3-14.2
	BETA βCyg ⊕ DOUBLE 3.0 & 5.1

*FIND VARIABLE STARS WITH "PROBLICOM" (P.132).

For DAYLIGHT SAVING TIMES see note on page 10.

15 JUL. MIDNIGHT
1 AUG. 11 P.M.
15 AUG. 10 P.M.
1 SEP. 9 P.M.
15 SEP. 8 P.M.

Glorious summer nights. The sky ablaze with myriad stars. The Milky Way waiting to be rediscovered, childhood memories to be recalled. What's out there? Where did we come from? Where are we going? Are we alone? How far is a galaxy? How long is eternity? The stars await.

The constellation of Cygnus the Swan is also known as the Northern Cross. The long upright and the shorter transverse beam form the classic emblem, with Alpha (**α**) Cygni, the brightest star, at the head and Gamma (**γ**) Cygni marking the point of intersection. Beta (**β**) Cygni marks the foot of the cruciform sign.

For the pictogram of the swan the direction is reversed. Beta (**β**) Cyg, the dazzling blue-and-gold double star called Albireo, represents the head. Alpha (**α**) Cyg, bright Deneb, forms the tail of the long-necked swan, and the cross arms have taken wing, extending beyond the width of the Milky Way along which the noble bird is gliding south.

Under dark skies and when viewed near the zenith, this is a starfield of rare beauty. It abounds with variable stars and is a region preferred by astrophotographers who conduct searches for novae, those explosive outbursts of radiant energy which increase the luminosity of stars by hundreds of thousands of times (page 120).

Cataclysmic stellar events can happen at any time, and there is no telling in advance just when or where a star might turn into a nova (*nova*-new). The Milky Way with its countless stars improves one's chances. Cygnus is "Novaland."

Novas (or novae) were often discovered by amateur stargazers who knew the sky very well and could immediately spot an extra

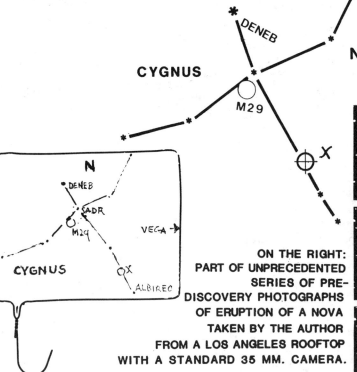

CYGNUS

DENEB

M29

NOVA CYGNI 1975
28-29 AUGUST
SEE PAGE 84

ON THE RIGHT:
PART OF UNPRECEDENTED
SERIES OF PRE-
DISCOVERY PHOTOGRAPHS
OF ERUPTION OF A NOVA
TAKEN BY THE AUTHOR
FROM A LOS ANGELES ROOFTOP
WITH A STANDARD 35 MM. CAMERA.

star, one that had not been there the night before. Such hard-earned knowledge is no longer necessary to make important discoveries. All that you need is a pair of identical photographs of a starfield. Why don't you center your camera on Vega tonight and record the first of the pair? Thereafter any photograph which you take of Cygnus — next week, next month or even next year —can be "blinked" (compared optically) with your first slide by using a simple home-built Problicom (pages 132-33). If there is anything new in the starfield or if a variable has brightened or dimmed, you will spot it immediately.

One of the brightest variables, Chi (**X**) Cygni, can thus be found near the neck of the swan. It ranges from naked-eye magnitude 3.3 to magnitude 14 and back again over a period of 407 days. There are dozens more variables to get you started.

It was during such a photographic survey that the author unknowingly recorded the explosion of a nova in Cygnus from a city rooftop with a standard 35mm camera.

GALACTIC CLUSTER

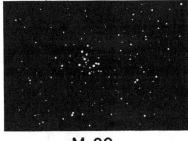

M 29

EXPOSURES: 12 MINS.

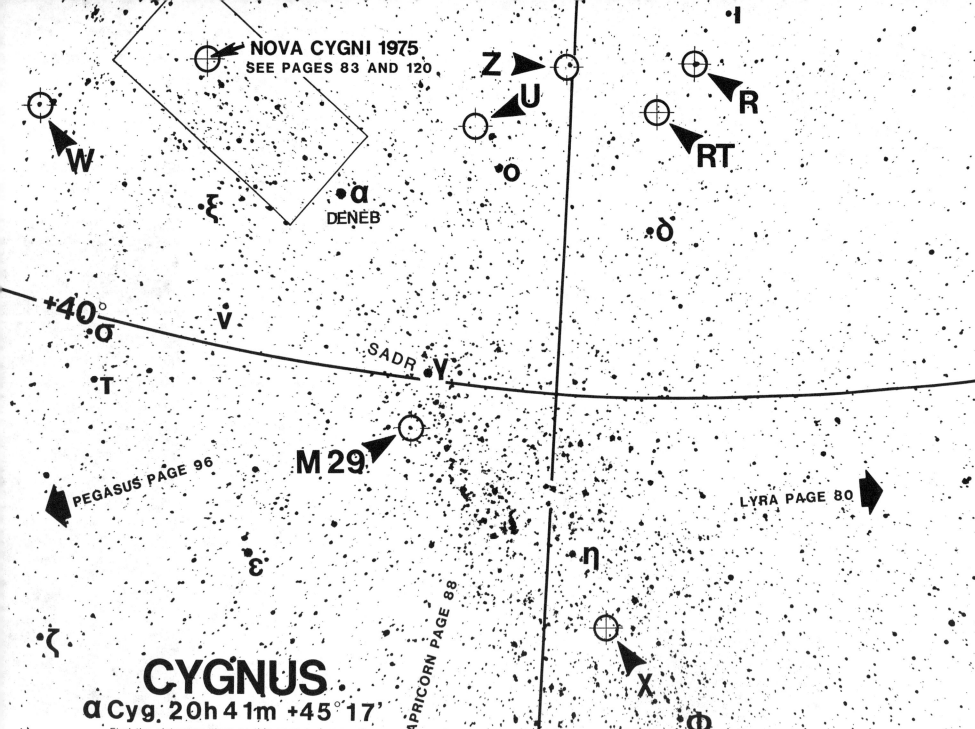

NOVA CYGNI 1975
SEE PAGES 83 AND 120

W

Z
U

I

R
RT

o

α
DENEB

ξ

δ

+40°
σ

ν

SADR γ

T

M 29

PEGASUS PAGE 96

LYRA PAGE 80

η

ε

CAPRICORN PAGE 88

ζ

X

CYGNUS
α Cyg 20h 41m +45°17'

φ

20h

β

Find the pictogram stars on this page and connect
them with ruled colored lines. You can also create your
own ideograms with red markings which will filter out
and disappear under a red flashlight (page 135).

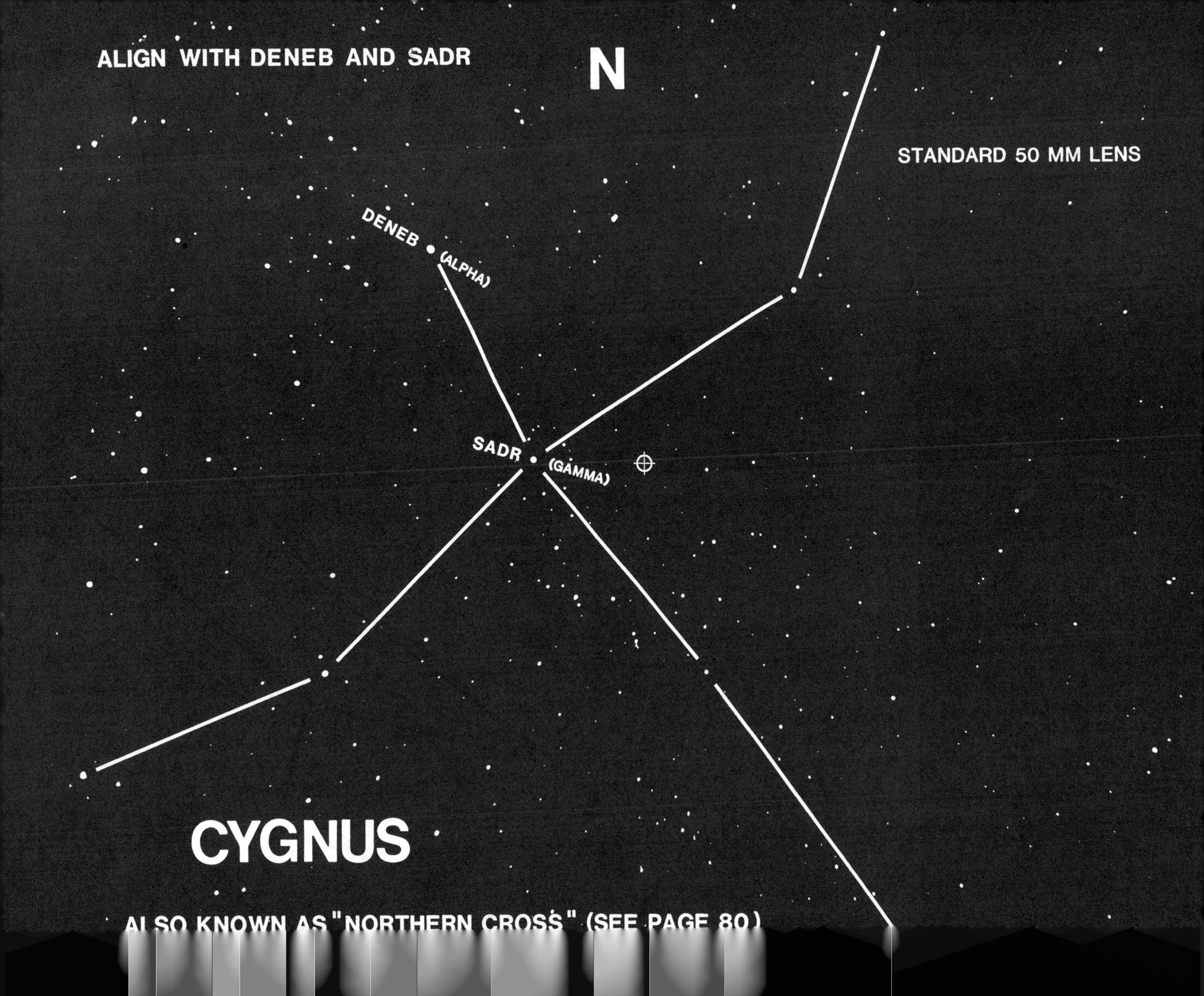

ALIGN WITH DENEB AND SADR

N

STANDARD 50 MM LENS

DENEB ● (ALPHA)

SADR ● (GAMMA) ⊕

CYGNUS

ALSO KNOWN AS "NORTHERN CROSS" (SEE PAGE 80)

L.Y.	BRIGHTEST STARS	MA
1,000	α_1 **ALGEDI** — THE GOAT (AR)	4.2
100	α_2 ALPHA	3.6
100	β **DABIH** — SLAUGHTERER'S STAR (AR) — BETA	3.1
100	γ **NASHIRA** — GOOD TIDINGS STAR (AR) — GAMMA	3.8
49	δ **DENEB ALGEDI** — TAIL OF THE GOAT (AR) — DELTA	2.9

THE Goat

CAP
Capricorni

CAPRICORN

Objects of Interest

L.Y.		
40,000	**M 30** ⊕ Globular cluster	
	T Cap ⊕ VARIABLE 269 DAYS 8.4–14.3	
	V Cap ⊕ VARIABLE 275 DAYS 8.2–14.0	
	RR Cap ⊕ VARIABLE 277 DAYS 7.8–14.6	
	RS Cap ⊕ VARIABLE 340 DAYS 8.3–10.3	
	ALPHA **α Cap** ∞ ⊕ DBL. DOUBLE 3.6 & 4.2	
	α^1 ⊕ 5.9 & 9.0 α^2 ⊕ 3.7 & 10.6	
	SIGMA **σ Cap** ⊕ DOUBLE 5.5 & 10.0	

NORTH STAR

SKY EQUATOR

+90° +80° +70° +60° +50° +40° +30° +20° +10° 0° -10° -20° -30° -40° -50°

+20°

HORIZON LINE

NORTH — CAPRICORN — SOUTH

EQUATOR

EARTH AXIS

For DAYLIGHT SAVING TIMES see note on page 10.

1 AUG. MIDNIGHT
15 AUG. 11 P.M.
1 SEP. 10 P.M.
15 SEP. 9 P.M.
1 OCT. 8 P.M.

In autumn it may be worth finding your own north-south-oriented bench in a park to use for a flatform when it gets colder and damp. Let it become your own observatory. Be sure you pick a site away from street lights and one which affords you four or five overhead "melon slices" of unobstructed seeing.

CAPRICORN

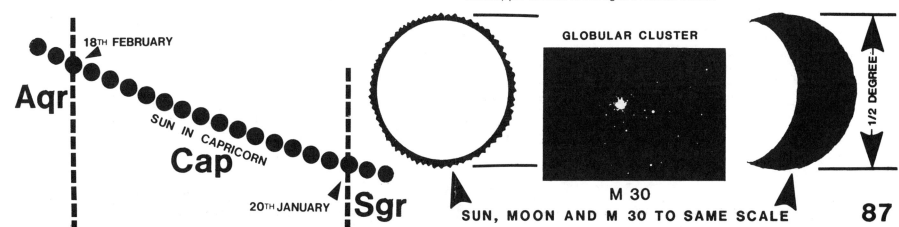

The constellation of Capricorn was known in ancient Babylon. This land, which lay in the Tigris-Euphrates valley in western Asia, shares with Egypt the distinction of being one of the earliest centers of civilization. The legends which surround the "Goatfish" starfield have their origins in Greek mythology, where different stories were invented to explain the unlikely creation of a goat with the tail of a fish.

Capricorn is a constellation of the Zodiac. The sun, having left the centaur of Sagittarius — a horse with the head of an archer — will not be too surprised to find itself once again in mixed company as it moves across the starfield of the "Sea-Goat" from January 20 to February 18. Not only is it difficult to recognize the pictogram of a creature of mythological imagination, but the stars of Capricorn are not among the brightest.

Algedi is a naked-eye double. The fainter of the pair is Alpha (α) 1 Capricorni, with a magnitude of 4.2. The brighter, Alpha (α) 2 Cap, has a magnitude of 3.6. The two stars are 376″ (arc-seconds) apart, which (divide by 60) equals more than 6′ (arc-minutes). This amounts to one tenth of a degree. (There are 60 arc-minutes to one degree.)

It may help to get a better grasp on scale and distance when we note that the globular cluster M30 (pictured below) has an apparent diameter of 6 arc-minutes. This means that in the sky it occupies an area which seems to be about as big around as the shape of Alpha (α) 1 and 2 Cap blending into each other in the photograph on the next page. (For comparison, the moon subtends an angle of 30 arc-minutes, or half of one degree in the sky.)

The rapture of the depths in space can be felt when we realize that Alpha (α) 2 Cap lies 100 light-years away from us, while Alpha (α) 1 Cap lies ten times as far. Both distances seem close when compared with the 40,000 light-years to faint M30, a globular cluster with thousands of stars.

When you view the dim glow of the globular in Capricorn through a telescope or see its stars etched into a photograph, you will be looking at 40,000-year-old light. The *time* it took for its photons to travel through space at a speed of 186,000 miles (300,000 kilometers) per second is the ages of modern man.

18TH FEBRUARY

Aqr

SUN IN CAPRICORN

Cap

20TH JANUARY

Sgr

GLOBULAR CLUSTER

M 30

1/2 DEGREE

SUN, MOON AND M 30 TO SAME SCALE

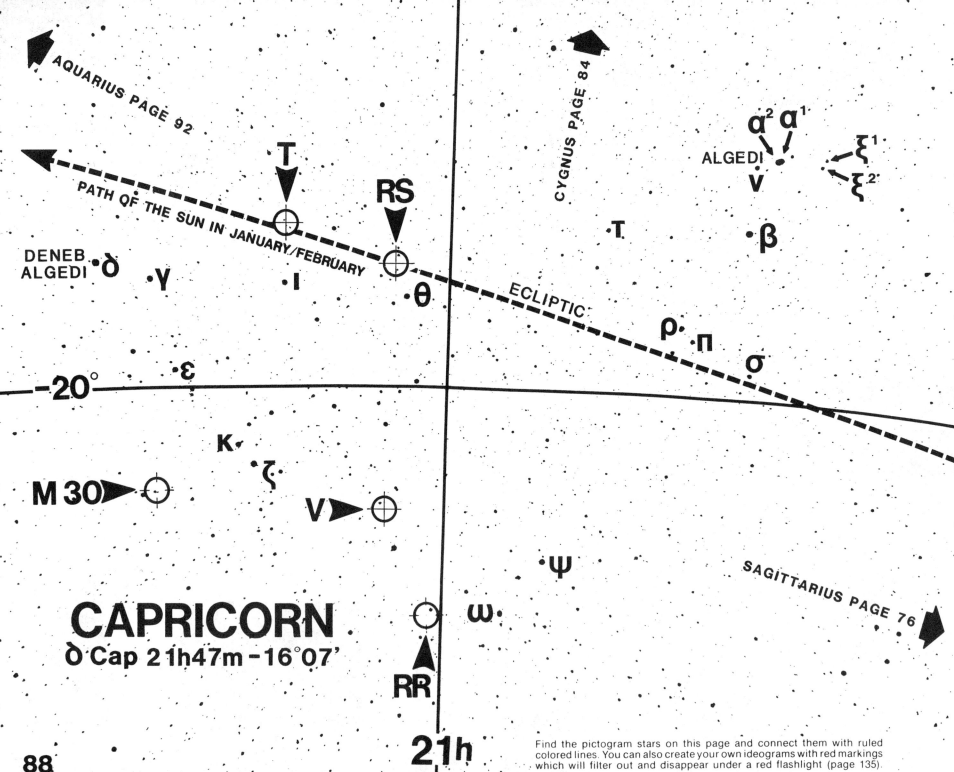

AQUARIUS PAGE 92

CYGNUS PAGE 84

α² α¹

ALGEDI

ξ¹

ξ²

ν

τ

•β

T

PATH OF THE SUN IN JANUARY/FEBRUARY

RS

DENEB
ALGEDI •δ

• γ

• ι

θ

ECLIPTIC

ρ •π

σ

−20°

ε

κ •

ζ

M 30

V

ψ

SAGITTARIUS PAGE 76

CAPRICORN
δ Cap 21h47m −16°07'

ω •

RR

21h

Find the pictogram stars on this page and connect them with ruled
colored lines. You can also create your own ideograms with red markings
which will filter out and disappear under a red flashlight (page 135).

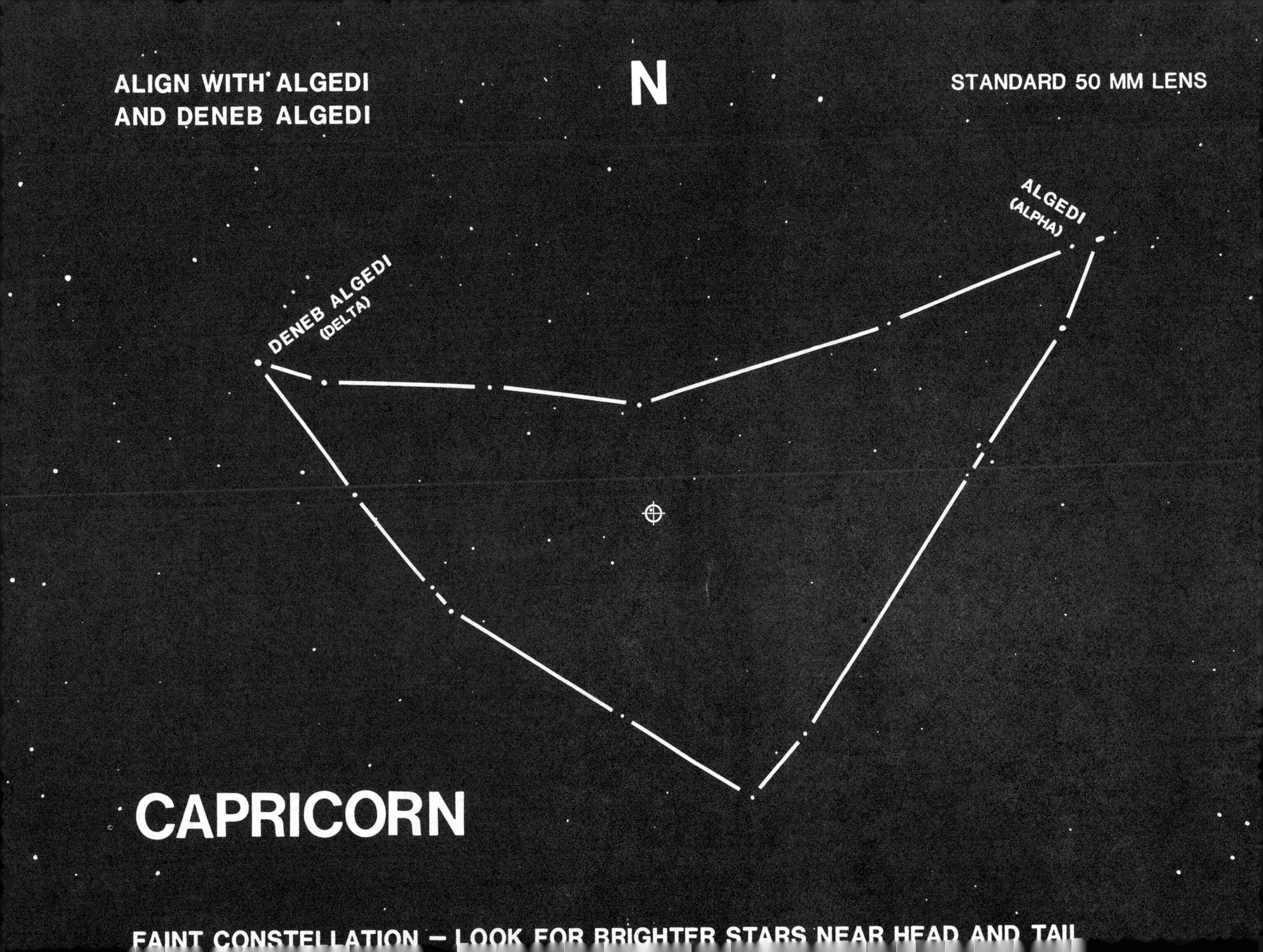

ALIGN WITH ALGEDI
AND DENEB ALGEDI
N
STANDARD 50 MM LENS
ALGEDI
(ALPHA)
DENEB ALGEDI
(DELTA)
CAPRICORN
FAINT CONSTELLATION – LOOK FOR BRIGHTER STARS NEAR HEAD AND TAIL

L.Y.		BRIGHTEST STARS	MAG
950	α ALPHA	SADALMELIK LUCKSTAR OF THE KING (AR)	3.0
980	β BETA	SADALSUD LUCKIEST OF LUCKSTARS (AR)	2.9
95	γ GAMMA	SADALACHBIA LUCKSTAR OF TENTS (AR)	3.8
98	δ DELTA	SKAT SHIN OF WATERBEARER (AR)	3.3
170	ε EPSILON	AL BALI SWALLOWER'S LUCKSTAR (AR)	3.7
75	ζ ZETA	ZETA AQR	3.6

WASSERMANN

THE Waterbearer

AQR
Aquarii

AQUARIUS

Objects of Interest

L.Y.		
50,000	M 2 ⊕	Globular cluster
85	NGC 7293 ⊕ VERY FAINT	Sunflower nebula

THIS NEBULA HAS A DIAMETER OF 12'X 16'(ABOUT HALF OF THE MOON). THE "SURFACE BRIGHTNESS" IS VERY LOW.

S Aqr ∞ ⊕	VARIABLE 278 DAYS	7.6-15.0
X Aqr ∞ ⊕	VARIABLE 311 DAYS	7.5-14.8
RY Aqr ⊕	VARIABLE 2 DAYS	8.8-10.1
BETA β Aqr ⊕	DOUBLE	3.0 & 11.0
ZETA ζ Aqr ⊕	DOUBLE	4.5 & 4.5

NORTH STAR

SKY EQUATOR

+90 +80 +70 +60 +50 +40 +30 +20 +10 0 -10 -20 -30 -40 -50

-10

HORIZON LINE

NORTH AQUARIUS SOUTH

EQUATOR EARTH AXIS

For DAYLIGHT SAVING TIMES see note on page 10.

1 SEP. MIDNIGHT
15 SEP. 11 P.M.
1 OCT. 10 P.M.
15 OCT. 9 P.M.
1 NOV. 8 P.M.

Be on the lookout for Orionid meteors from October 17 to the 26th. Your best night is October 20. Orionids are believed to come from Halley's Comet. When the earth in its journey around the sun crosses earlier orbits of the historic visitor, we coast through some of Halley's cometary celestial debris.

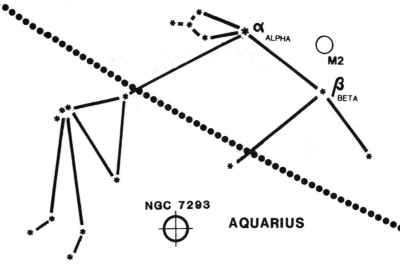

AQUARIUS

The constellation of Aquarius and the two starfields flanking it in the Zodiac are all associated with water. Neighboring Capricorn has at least the tail of a fish to relate it to the liquid element. Adjacent Pisces offers not one but two fishes in their natural environment. This is the region of the sky which was called "water-filled" by the ancients. Babylonians referred to it as the "sea." Other constellations, less well known, lie nearby. Their names tell the story: the Dolphin, the Whale, the Southern Fish.

The sun is in Aquarius from February 17 to March 13. The pictogram vaguely portrays a water bearer in the form of a man or a boy pouring water from a two-handled wine jar. In ancient Egypt the starfield was said to be known as the "home of the flowing waters."

Priests related the setting of this starfield to the annual flooding of the Nile. But then the rising of bright Sirius (page 43) was also associated with this event, which is not surprising when one considers the importance of irrigation to agriculture in arid lands. The earliest astrological signs for Aquarius look just like the Egyptian hieroglyph for water.

For the Arabs, the stars in this region held some association with "good fortune within the tents." Four of the principal stars here have some connotation with luck: Sadalmelik, Sadalsud, Sadalachbia and Al Bali.

M2 is a beautiful globular cluster half an R.A. hour to the right of Alpha (α) Aquarii and 5° north of Beta (β) Aqr. It appears as a 7th-magnitude star and lies at a distance of some 50,000 light-years from us.

There is a very beautiful, but faint "planetary nebula" (page 79) which Messier never saw even though its diameter is about half that of the moon (15 arc-minutes). It is called the Double Helix. Because it has a very low surface brightness and is very diffuse, you might only catch sight of a large circular hazy spot when you scan the exact area from a dark location with binoculars or a telescope at very low magnification.

As the photograph below indicates, even the pale light of rarefied gas in a nebula can be collected with a camera and will accumulate on film photon by photon, as in the case of the Sunflower nebula.

THE DOUBLE HELIX IS ALSO KNOWN

AS THE "SUNFLOWER NEBULA"

GLOBULAR CLUSTER

M 2

PLANETARY NEBULA

NGC 7293
DOUBLE HELIX

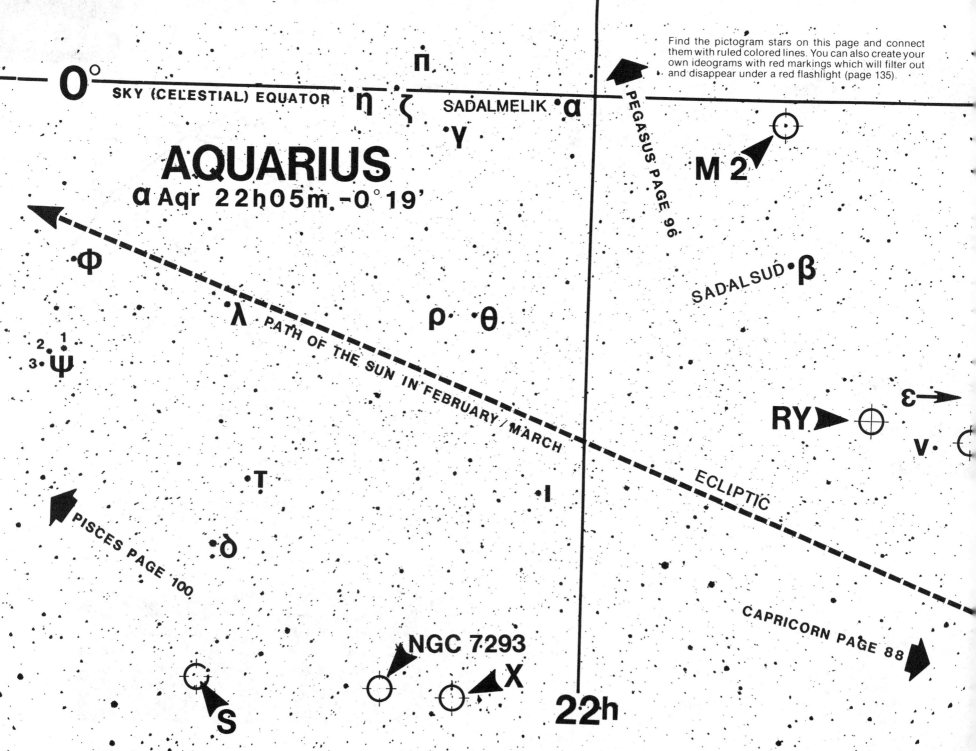

0°

SKY (CELESTIAL) EQUATOR   η̇ ζ   ṅ   π̇   SADALMELIK   α

γ

AQUARIUS
α Aqr 22h05m -0°19'

PEGASUS PAGE 96

Find the pictogram stars on this page and connect them with ruled colored lines. You can also create your own ideograms with red markings which will filter out and disappear under a red flashlight (page 135).

M 2

φ

λ   PATH OF THE SUN IN FEBRUARY / MARCH

SADALSUD • β

ρ • θ

1
2 • •
3 • ψ

RY ► ⊕   ε→

ν • ⊕

T

ι

ECLIPTIC

• δ

PISCES PAGE 100

CAPRICORN PAGE 88

NGC 7293

X

22h

S

WIDE-ANGLE LENS PHOTOGRAPHS TO CAPTURE ENTIRE CONSTELLATION • (SEE NOTE PAGE 57)

ALIGN WITH SADALMELIK
AND SADALSUD

WIDE-ANGLE 35 MM LENS

N

AQUARIUS

SADALMELIK
(ALPHA)

SADALSUD
(BETA)

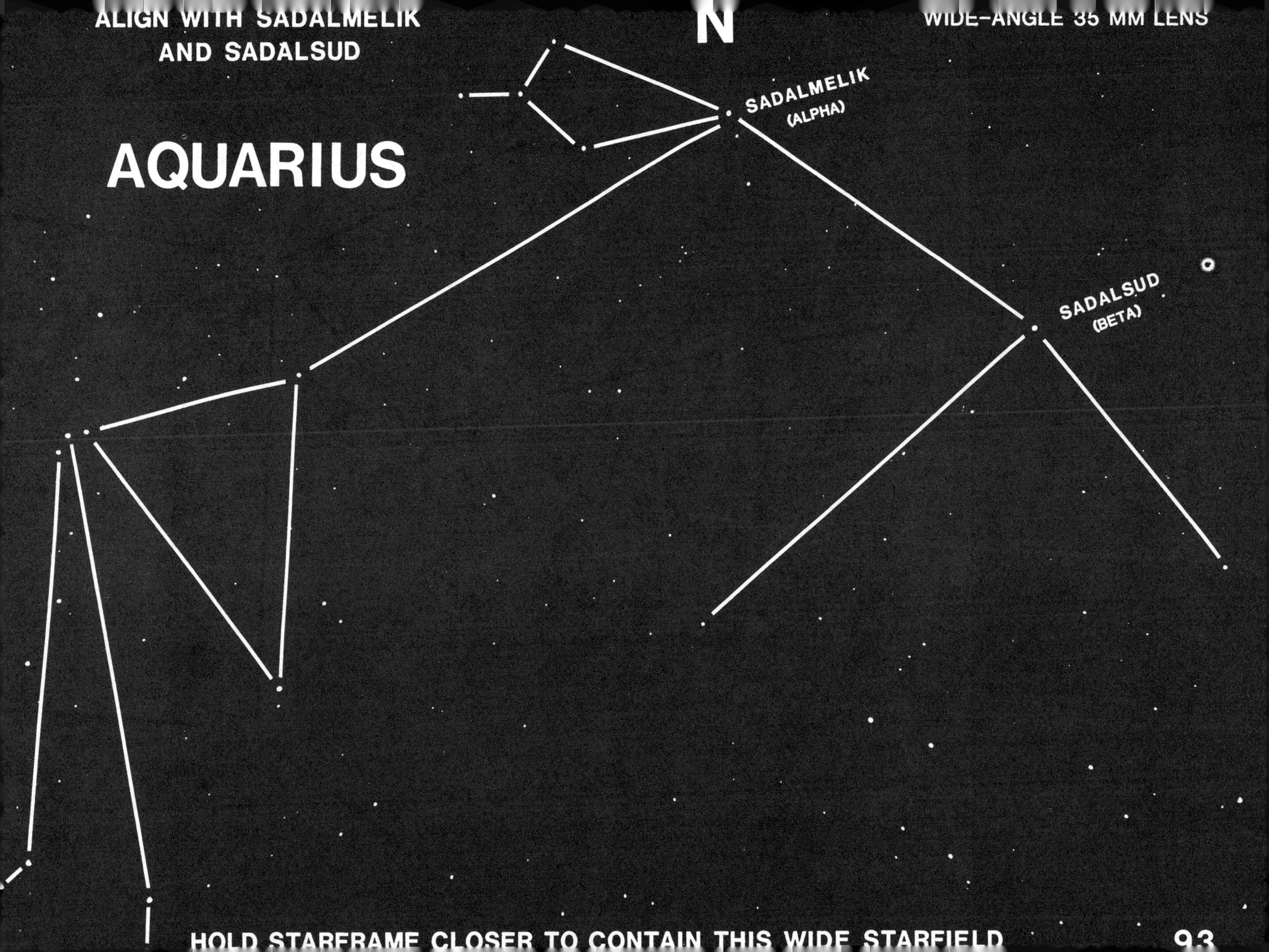

HOLD STARFRAME CLOSER TO CONTAIN THIS WIDE STARFIELD

93

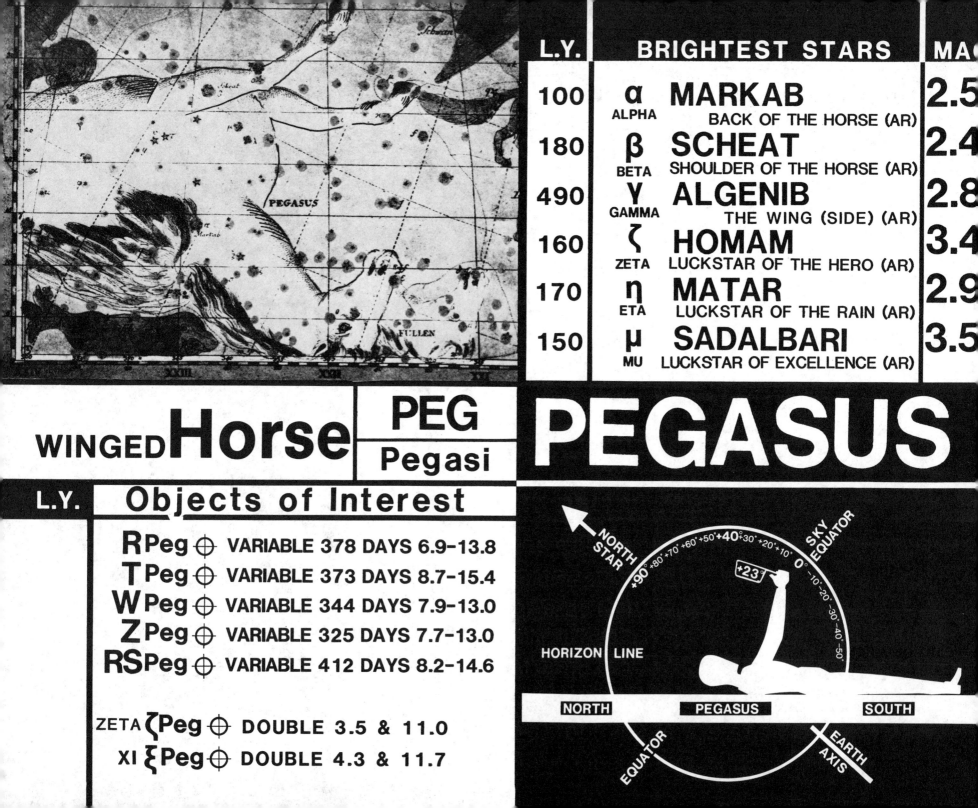

L.Y.	BRIGHTEST STARS		MAG
100	α ALPHA	**MARKAB** BACK OF THE HORSE (AR)	2.5
180	β BETA	**SCHEAT** SHOULDER OF THE HORSE (AR)	2.4
490	γ GAMMA	**ALGENIB** THE WING (SIDE) (AR)	2.8
160	ζ ZETA	**HOMAM** LUCKSTAR OF THE HERO (AR)	3.4
170	η ETA	**MATAR** LUCKSTAR OF THE RAIN (AR)	2.9
150	μ MU	**SADALBARI** LUCKSTAR OF EXCELLENCE (AR)	3.5

WINGED Horse | PEG / Pegasi | PEGASUS

L.Y.	Objects of Interest	
	R Peg ⊕	VARIABLE 378 DAYS 6.9-13.8
	T Peg ⊕	VARIABLE 373 DAYS 8.7-15.4
	W Peg ⊕	VARIABLE 344 DAYS 7.9-13.0
	Z Peg ⊕	VARIABLE 325 DAYS 7.7-13.0
	RS Peg ⊕	VARIABLE 412 DAYS 8.2-14.6
	ZETA ζ **Peg** ⊕	DOUBLE 3.5 & 11.0
	XI ξ **Peg** ⊕	DOUBLE 4.3 & 11.7

NORTH STAR

+90° +80° +70° +60° +50° +40° +30° +20° +10° 0° -10° -20° -30° -40° -50°

SKY EQUATOR

+23°

HORIZON LINE

NORTH | PEGASUS | SOUTH

EQUATOR

EARTH AXIS

For DAYLIGHT SAVING TIMES see note on page 10.

15 SEP. MIDNIGHT
1 OCT. 11 P.M.
15 OCT. 10 P.M.
1 NOV. 9 P.M.
15 NOV. 8 P.M.

Subscribe or make a gift of a subscription to one of the many astronomical/science magazines which are available; some are listed under "Resources" (page 142). Do it early, in time for the holidays. Then enjoy some winter armchair astronomy to prepare for the glories of nights to come.

The great "square" of Pegasus is actually a rectangle defined by four 2nd-magnitude stars. Even from a dark location not too many other fainter stars lie in the area, which is why the geometric form stands out. It is somewhat puzzling to learn that the rectangle constitutes the body of a horse with golden wings. It flies seemingly upside down in the night sky. The Bode atlas illustration will have to fill in the details here, because the star pictogram seems of little help.

The square of Pegasus can be made to serve a useful purpose. It permits a better understanding of directions and dimensions in the sky. On its long east-west "horizontal" axis, the sides are parallel to the celestial equator and both are a little over one R.A. hour wide. The shorter, north-south "vertical" sides subtend (stretch out over the distance of) fourteen degrees of declination.

If we visualize the square of Pegasus superimposed on one of the twenty-four melon-peel slices of the celestial sphere, then the entire bottom line, extending from Alpha (α) Pegasi to Gamma (γ) Peg lies at +15° Dec. The upper horizontal lies at +28°. Both are parallel to each other and to the celestial equator at 0°. The vertical lines of the square are a little different. In the north-south direction, lines connecting the poles, while straight in themselves, are never parallel to each other. They are joined at the North Celestial Pole, then split until they are widest apart at the celestial equator. Soon they curve back in until they meet again at the South Celestial Pole. For this reason, an "hour of right ascension" covers a shorter distance on the celestial sphere the farther we move north or south, away from the celestial equator.

A shape like the square of Pegasus, about an hour wide in right ascension, when moved nearer to the poles will retain its height in declination, but its apparent width would shrink considerably, as shown. Yet, the squares illustrated on the right are all "one hour wide by fifteen degrees high." Four Pegasus rectangles should just about fill a starframe. This means that on the scale of this book each page or starframe is about two R.A. hours wide by thirty degrees high near the celestial equator.

Clockwise, the corners of the rectangle in Pegasus are marked by the stars Scheat, Markab, Algenib and Alpheratz. Of these the first three belong to Pegasus. The fourth, in the top left corner, has a dual personality. When regarded as Delta (δ) Pegasi its name denotes the "navel of the horse." For the Andromeda pictogram, it becomes Alpha (α) Andromedae and turns into a woman's head.

SCHEAT

ALPHERAZ

W

PEGASUS

ALGENIB

MARKAB
α ALPHA

+90°
+80°
+70°
HR x 15
+60°
HR x 15
+50°
+40°
+30°
+30 DEGREES
Pegasus Square
+20°
HR x 15
+10°
0°
0 HRS   23 HRS IN R.A.
-10°
FLATTENED
"MELONSLICE"
-20°
-30°
-40°
-50°
-60°
-70°
-80°
-90°

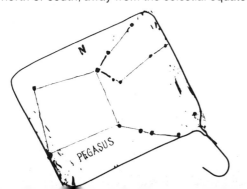

N

PEGASUS

95

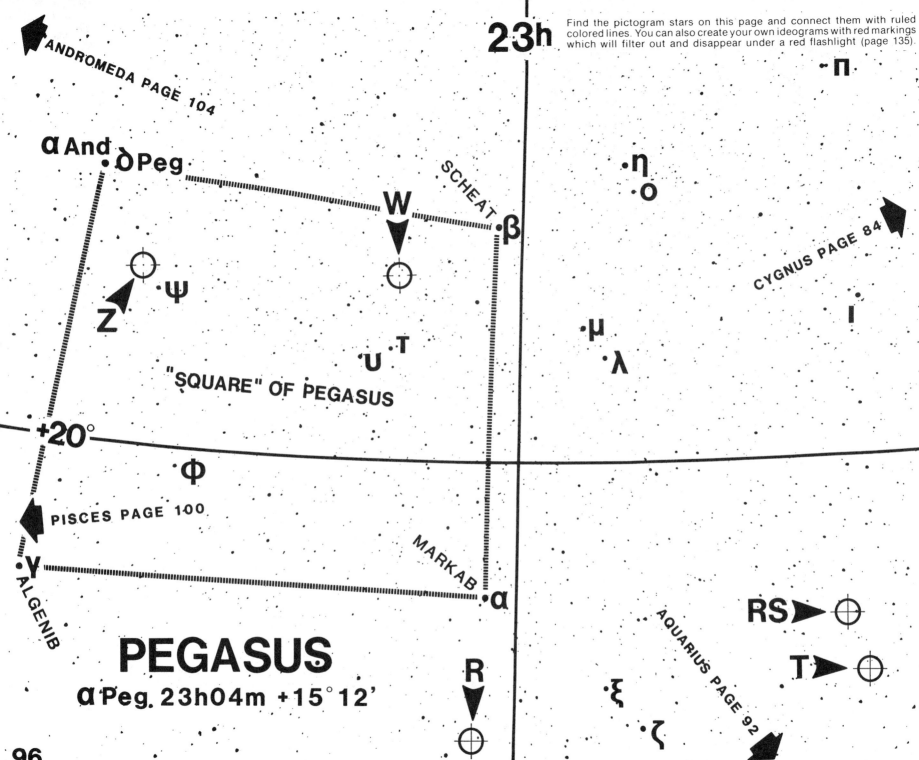

Find the pictogram stars on this page and connect them with ruled colored lines. You can also create your own ideograms with red markings which will filter out and disappear under a red flashlight (page 135).

ANDROMEDA PAGE 104

α And   δ Peg

SCHEAT

W

β

η
o

CYGNUS PAGE 84

ψ

Z

μ
λ

I

υ τ

"SQUARE" OF PEGASUS

+20°

Φ

PISCES PAGE 100

MARKAB

γ

ALGENIB

α

PEGASUS
α Peg. 23h04m +15°12'

R

RS

T

ξ
ζ

AQUARIUS PAGE 92

ALIGN WITH MARKAB AND SCHEAT

STANDARD 50 MM LENS

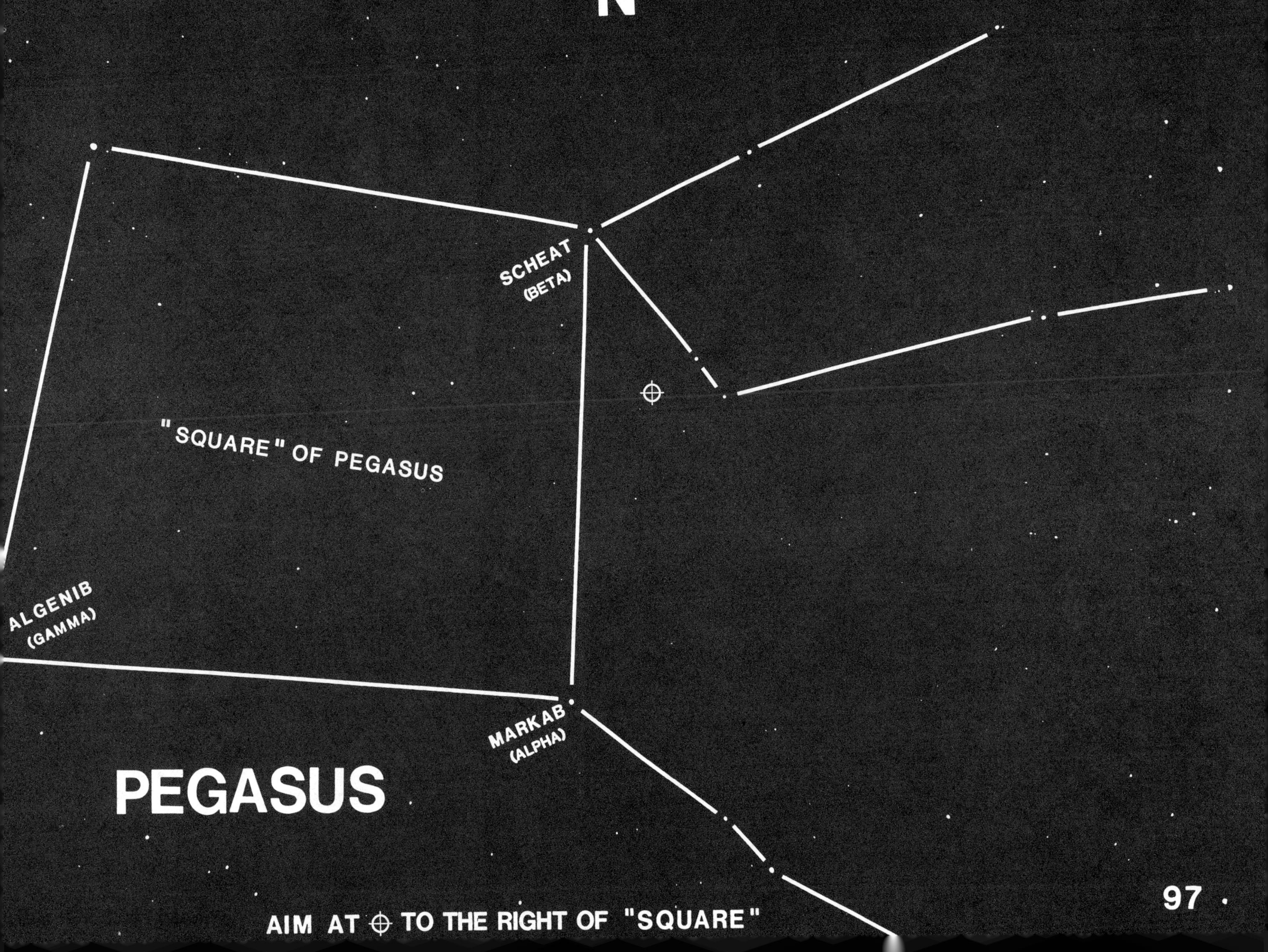
N
SCHEAT
(BETA)
"SQUARE" OF PEGASUS
ALGENIB
(GAMMA)
MARKAB
(ALPHA)
PEGASUS

AIM AT ⌖ TO THE RIGHT OF "SQUARE"

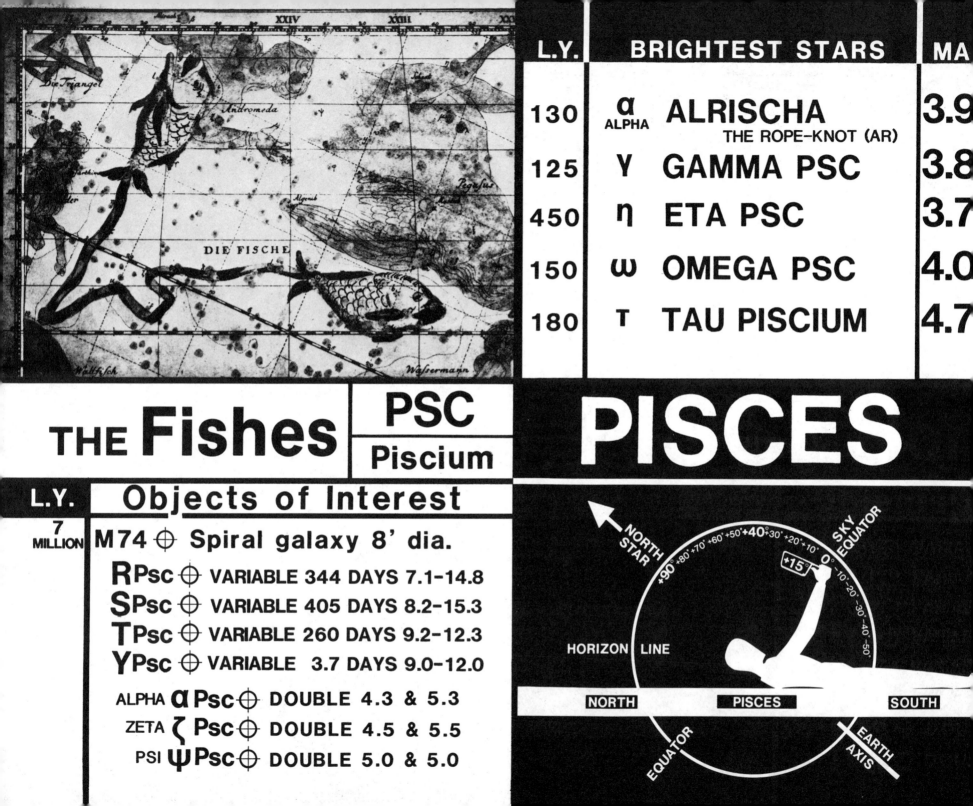

PISCES

THE Fishes | PSC / Piscium

Objects of Interest

L.Y.		
7 MILLION	M 74 ⊕	Spiral galaxy 8' dia.
	R Psc ⊕	VARIABLE 344 DAYS 7.1-14.8
	S Psc ⊕	VARIABLE 405 DAYS 8.2-15.3
	T Psc ⊕	VARIABLE 260 DAYS 9.2-12.3
	Y Psc ⊕	VARIABLE 3.7 DAYS 9.0-12.0
	ALPHA α Psc ⊕	DOUBLE 4.3 & 5.3
	ZETA ζ Psc ⊕	DOUBLE 4.5 & 5.5
	PSI ψ Psc ⊕	DOUBLE 5.0 & 5.0

NORTH STAR

SKY EQUATOR

+90° +80° +70° +60° +50° +40° +30° +20° +10° 0° -10° -20° -30° -40° -50°

+15°

HORIZON LINE

NORTH — PISCES — SOUTH

EQUATOR

EARTH AXIS

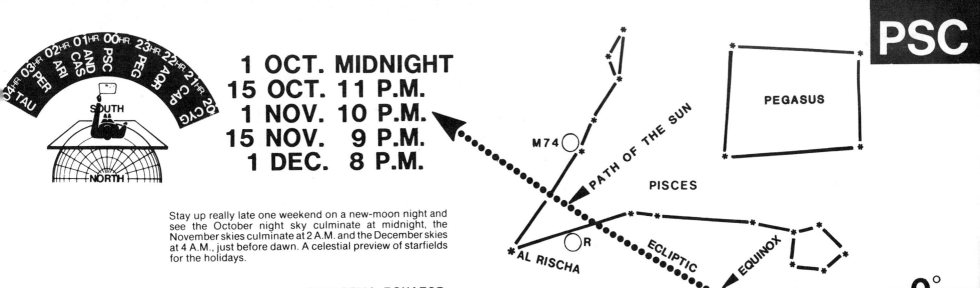

1 OCT. MIDNIGHT
15 OCT. 11 P.M.
1 NOV. 10 P.M.
15 NOV. 9 P.M.
1 DEC. 8 P.M.

Stay up really late one weekend on a new-moon night and see the October night sky culminate at midnight, the November skies culminate at 2 A.M. and the December skies at 4 A.M., just before dawn. A celestial preview of starfields for the holidays.

CELESTIAL EQUATOR — 0°

The constellation was named "Gemini Pisces" by the Romans because there are two fish. The "western" fish heads northward toward Andromeda, and the "eastern" fish back in the direction of Aquarius. Alrischa, "the Knot," is not a bright star but connects the pair, which together form a large L shape, cradling the square of Pegasus.

The sun enters Pisces on March 13 and remains there until April 19. During that time it crosses the equator from the southern to the northern celestial hemisphere on about March 22. That point in *time* is regarded as the beginning of spring, the Vernal Equinox, in the northern part of the globe. The equinox is also the particular *place* where the sun actually crosses the celestial equator in its diagonal climb. It may be remembered that this point marks the zero-hour position in R.A. from which the twenty-four hours in right ascension are counted. That is why the coordinates of the point of the Vernal Equinox are 00:00 hrs R.A. and 0° declination. Today that point is in Pisces.

Equi stands for "equal," and *nox* means night. At the time of equinox, nights and days are of equal lengths (pages 140-41). There is a second point of equinox in the celestial sphere. The position is 12:00 hrs R.A., 0° Dec. Its point in time marks the beginning of fall in the Northern Hemisphere. Nights and days are once again of equal lengths when the sun crosses the celestial equator on its southward journey in the constellation Virgo (pages 59-60).

The gravitational pull of the sun and the moon on the earth creates a very gentle wobble in the north-south tilt of the earth. It affects the direction in which our polar axis is pointed and the angle of the celestial equator in relation to the path of the sun and the Zodiac. This effect is called "precession," and it has been compared to the wobble which can be observed in spinning tops. For earth, each jiggle takes 26,000 years to complete. Precession very gradually changes the apparent positions of the stars in relation to us. The "first point in Aries" used to be the point of the spring equinox when early records were kept some two thousand years ago. Since then it has moved into Pisces, and it will, about six hundred years from now, drift into Aquarius. This was not known to the ancients. We honor early observers by still marking the points of the equinoxes with the signs of the constellations in which they were first recorded: the symbol of Aries in spring and the mark of Libra in autumn.

SPIRAL GALAXY

M 74   8' DIAMETER

99

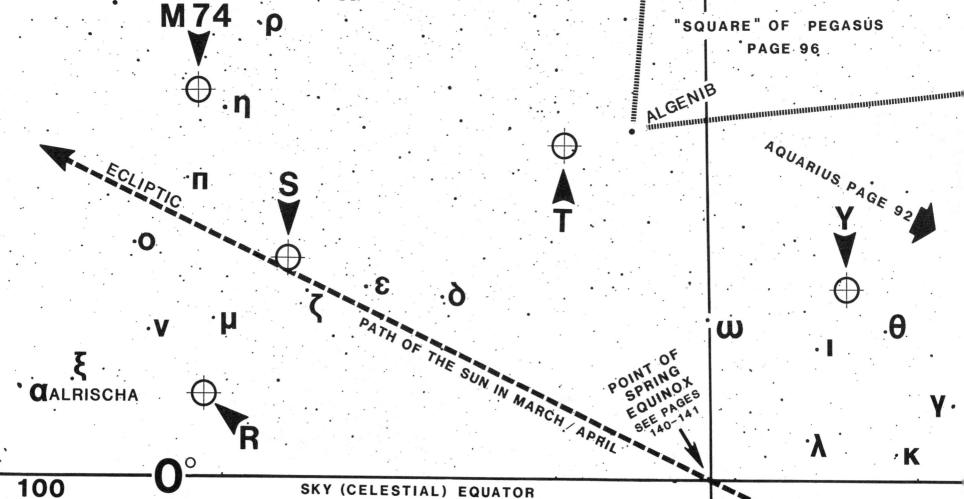

Find the pictogram stars on this page and connect them with ruled colored lines. You can also create your own ideograms with red markings which will filter out and disappear under a red flashlight (page 135).

τ σ

0h

U

ANDROMEDA PAGE 104

PISCES
α Psc 2h02m +2°46'

φ

ARIES PAGE 112

χ ψ

M74 ρ

"SQUARE" OF PEGASUS
PAGE 96

η

ALGENIB

AQUARIUS PAGE 92

ECLIPTIC

π

S

Y

ο

T

ω

θ

ε δ

ι

ζ

PATH OF THE SUN IN MARCH / APRIL

ν μ

ξ

POINT OF
SPRING
EQUINOX
SEE PAGES
140–141

γ

α ALRISCHA

R

λ κ

O°

SKY (CELESTIAL) EQUATOR

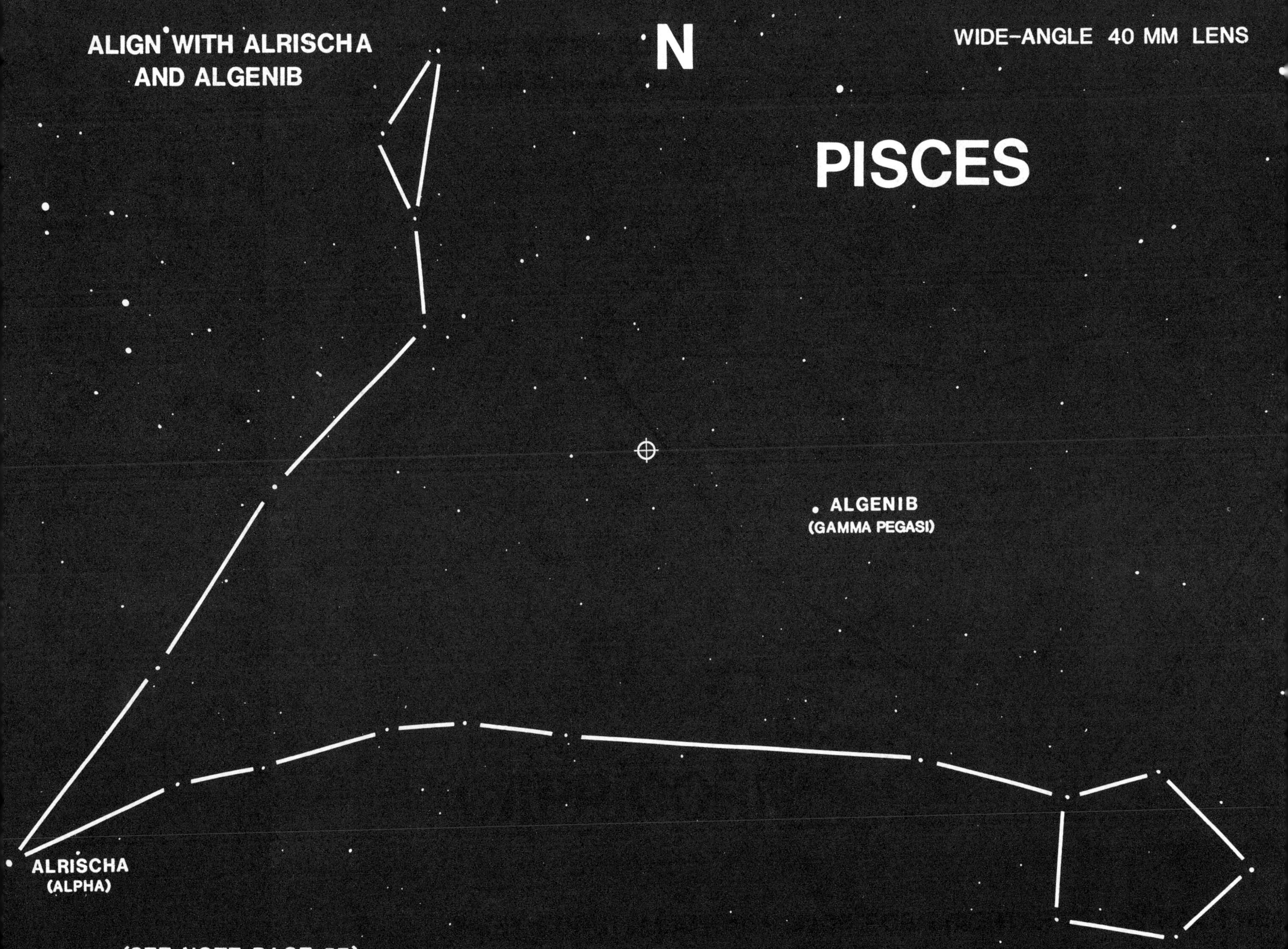

ALIGN WITH ALRISCHA
AND ALGENIB

WIDE-ANGLE 40 MM LENS

PISCES

ALGENIB
(GAMMA PEGASI)

ALRISCHA
(ALPHA)

(SEE NOTE PAGE 57)

HOLD STARFRAME CLOSER TO CONTAIN THIS LARGE CONSTELLATION

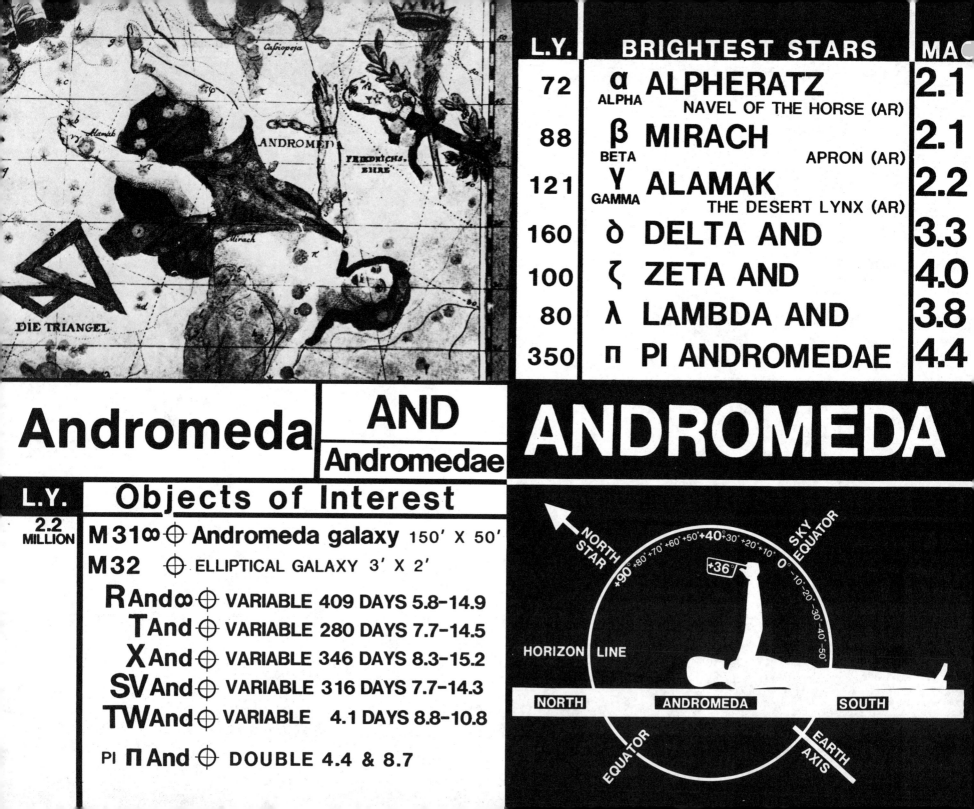

L.Y.	BRIGHTEST STARS	MAG
72	α ALPHA ALPHERATZ NAVEL OF THE HORSE (AR)	2.1
88	β BETA MIRACH APRON (AR)	2.1
121	γ GAMMA ALAMAK THE DESERT LYNX (AR)	2.2
160	δ DELTA AND	3.3
100	ζ ZETA AND	4.0
80	λ LAMBDA AND	3.8
350	π PI ANDROMEDAE	4.4

Andromeda | AND | ANDROMEDA
Andromedae

L.Y.	Objects of Interest
2.2 MILLION	M 31 ∞ ⊕ Andromeda galaxy 150' X 50'
	M 32 ⊕ ELLIPTICAL GALAXY 3' X 2'
	R And ∞ ⊕ VARIABLE 409 DAYS 5.8-14.9
	T And ⊕ VARIABLE 280 DAYS 7.7-14.5
	X And ⊕ VARIABLE 346 DAYS 8.3-15.2
	SV And ⊕ VARIABLE 316 DAYS 7.7-14.3
	TW And ⊕ VARIABLE 4.1 DAYS 8.8-10.8
	PI π And ⊕ DOUBLE 4.4 & 8.7

NORTH STAR

SKY EQUATOR

+90° +80° +70° +60° +50° +40° +30° +20° +10° 0° -10° -20° -30° -40° -50°

+36°

HORIZON LINE

NORTH   ANDROMEDA   SOUTH

EQUATOR

EARTH AXIS

1 OCT. MIDNIGHT
15 OCT. 11 P.M.
1 NOV. 10 P.M.
15 NOV. 9 P.M.
1 DEC. 8 P.M.

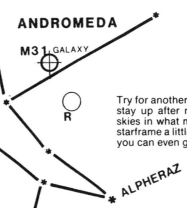

ANDROMEDA

M31 GALAXY

R

ALPHERAZ

Try for another "preview night." Select a new-moon date to stay up after midnight and observe cold January winter skies in what may yet be milder weather. If you hold your starframe a little to the left (east) of the straight-up position, you can even glimpse the February midnight skies.

This constellation is home to the most famous of the island universes, the magnificent Andromeda galaxy. Because it is a naked-eye object under dark skies, with a total visual magnitude of 4, it was dimly seen and reported at least as far back as a thousand years ago. At first it was referred to as a "little cloud," later as a "nebula." Messier assigned the number 31 to it and 32 to an adjoining companion.

Not until long-exposure photographs could be taken with large instruments were the first component stars revealed. This established M31 as a galaxy of countless stars rather than a cloud of glowing gas. Although most of its stars remain invisible, astronomers today estimate that the Andromeda galaxy consists of about 300 billion individual stars. Some of these are variables of the same kind we have in our own galaxy, whose changing luminosity permits scientists to determine astronomical distances rather accurately. On the basis of such observations we can calculate that M31 is approximately 2.2 million light-years away from us.

One of the reasons why the Andromeda system may hold a special fascination for observers is that it is a member of our "local group of galaxies" — a neighbor, so to speak. With an estimated diameter of some 200,000 light-years, it is a spiral about twice the size of our Milky Way galaxy. M31 has its own globular clusters, open clusters, novae and supernovae. It is quite conceivable that solar systems like our own exist within its vast expanse, suns with planets which, in turn, have moons of their own.

From the Andromeda perspective, our galaxy would appear not unlike the hazy patch which we see when we view M31 tonight. Because the speed of light is constant from any viewpoint, our galaxy would be seen as it was two million years ago, just as we observe an equally ancient Andromeda galaxy tonight. Even with the world's largest telescopes we can resolve only a limited number of the very brightest stars in M31. Thousands of millions of others will remain a mere glow in space. Similar optical giants trained on our Milky Way galaxy from a distance of 2 million light-years could not even detect our sun. It would be 200 times too faint to see.

M 31 ANDROMEDA GALAXY

M 32

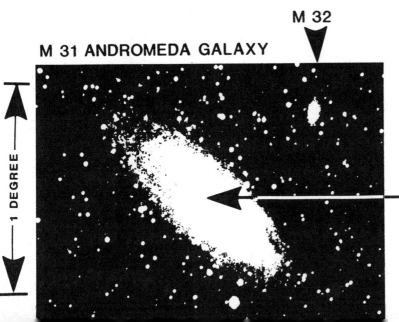

1 DEGREE

2,200,000 LIGHT-YEARS

OUR GALAXY

Φ

X ▶ ⊕

0h

Ψ • λ

PERSEUS PAGE 116

γ

K •

• I

M31
▼

T • U

ν •

M32

SV

μ

θ

R

ρ

σ

MIRACH • β

ARIES PAGE 112

TW

PEGASUS PAGE 96

π

+30°

δ

PISCES PAGE 100

ε

ALPHERATZ • α

ANDROMEDA
α And 0h08m +29°05'

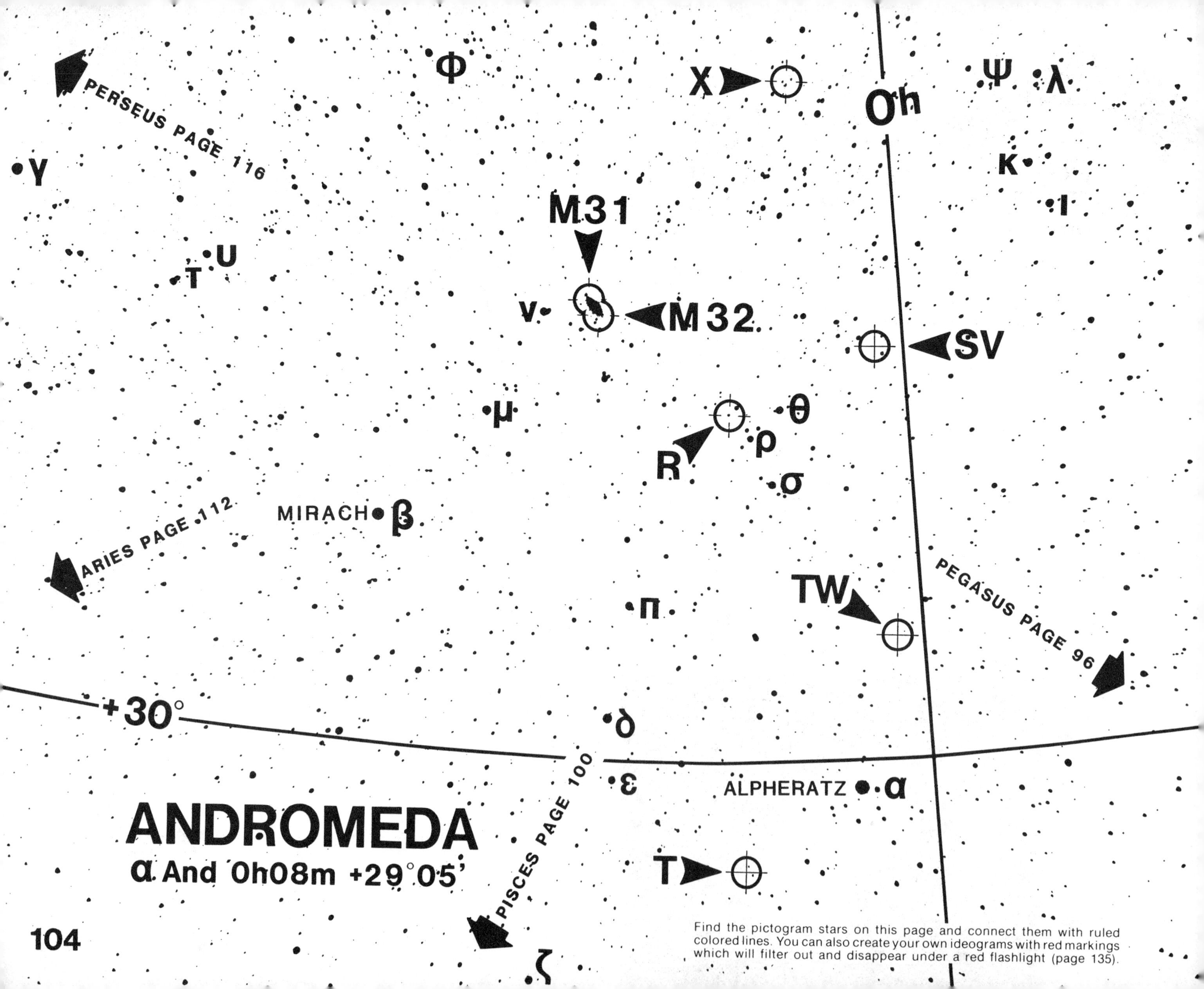

T ▶ ⊕

ζ

Find the pictogram stars on this page and connect them with ruled colored lines. You can also create your own ideograms with red markings which will filter out and disappear under a red flashlight (page 135).

104

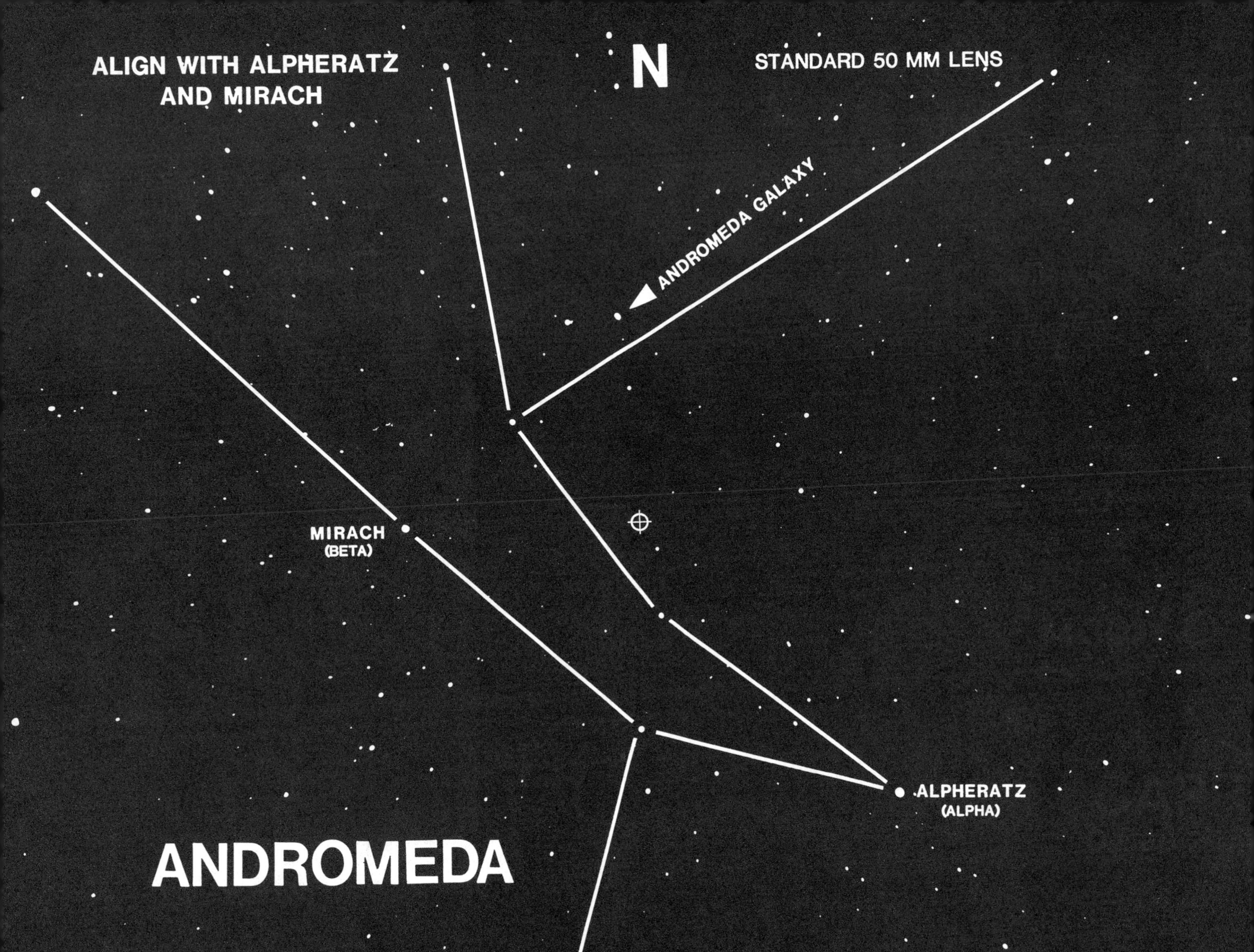

ALIGN WITH ALPHERATZ
AND MIRACH

N

STANDARD 50 MM LENS

ANDROMEDA GALAXY

MIRACH
(BETA)

ALPHERATZ
(ALPHA)

ANDROMEDA

THE ANDROMEDA GALAXY IS FAINT, BUT IS VISIBLE TO THE NAKED EYE

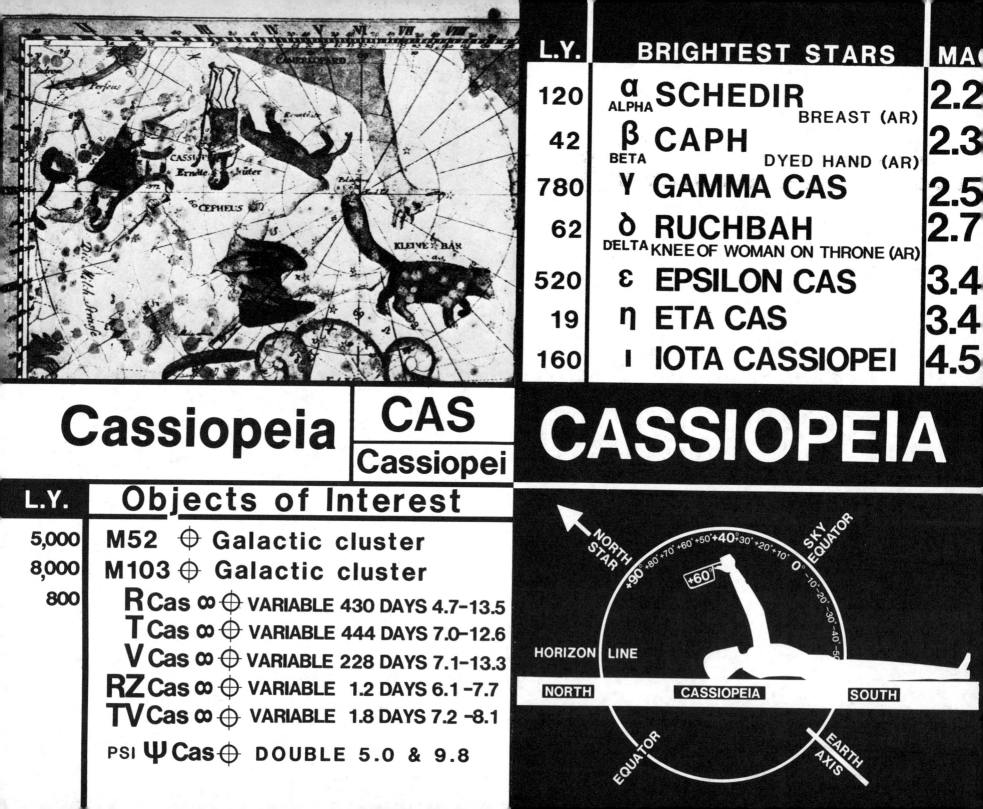

L.Y.	BRIGHTEST STARS		MAG
120	α ALPHA	SCHEDIR BREAST (AR)	2.2
42	β BETA	CAPH DYED HAND (AR)	2.3
780	γ	GAMMA CAS	2.5
62	δ DELTA	RUCHBAH KNEE OF WOMAN ON THRONE (AR)	2.7
520	ε	EPSILON CAS	3.4
19	η	ETA CAS	3.4
160	ι	IOTA CASSIOPEI	4.5

Cassiopeia

CAS
Cassiopei

CASSIOPEIA

Objects of Interest

L.Y.			
5,000	M52 ⊕	Galactic cluster	
8,000	M103 ⊕	Galactic cluster	
800	R Cas ∞ ⊕	VARIABLE 430 DAYS	4.7-13.5
	T Cas ∞ ⊕	VARIABLE 444 DAYS	7.0-12.6
	V Cas ∞ ⊕	VARIABLE 228 DAYS	7.1-13.3
	RZ Cas ∞ ⊕	VARIABLE 1.2 DAYS	6.1-7.7
	TV Cas ∞ ⊕	VARIABLE 1.8 DAYS	7.2-8.1
	PSI Ψ Cas ⊕	DOUBLE 5.0 & 9.8	

NORTH STAR

SKY EQUATOR

+90° +80° +70° +60° +50° +40° +30° +20° +10° 0° -10° -20° -30° -40° -50°

+60°

HORIZON LINE

NORTH — CASSIOPEIA — SOUTH

EQUATOR

EARTH AXIS

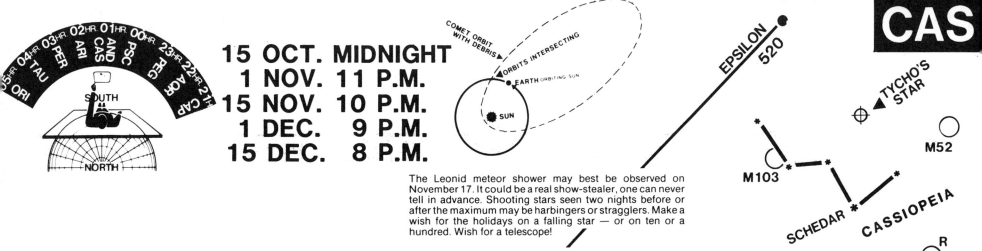

15 OCT. MIDNIGHT
1 NOV. 11 P.M.
15 NOV. 10 P.M.
1 DEC. 9 P.M.
15 DEC. 8 P.M.

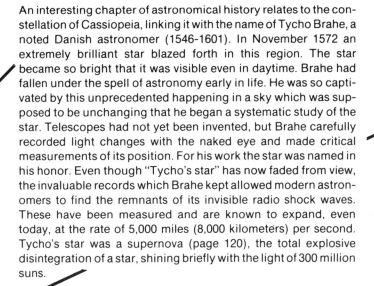

The Leonid meteor shower may best be observed on November 17. It could be a real show-stealer, one can never tell in advance. Shooting stars seen two nights before or after the maximum may be harbingers or stragglers. Make a wish for the holidays on a falling star — or on ten or a hundred. Wish for a telescope!

With its familiar zigzag of five stars, Cassiopeia, not far from Polaris, is an easy-to-find constellation. The bright stars range from a magnitude of 2.2 for Alpha (**α**) Cassiopei to 3.4 for Epsilon (**ε**) Cas. Together they form a W, seemingly all in the same plane. The diagram below, however, gives their different distances from us drawn to the same scale. They average about 305 light-years, which seems close when compared to how far it is to M52 and M103. These open clusters are twenty times as remote. When thought of in terms of other extremely faint galactic companions of M31, which happen to lie in the constellation Cassiopeia seven thousand times farther out, the immense depth of space begins to make itself felt. On the scale of our diagram such galaxies lie at a distance of 2.5 miles from this page.

There are several very bright variables in Cassiopeia. They include R Cas, which ranges from magnitude 4.7 to 13.5 and back over a period of 430 days; also the so-called "short-period" variables RZ Cas and TV Cas, which have lesser ranges of magnitude but vary with periods of just thirty and fifty hours (page 121).

An interesting chapter of astronomical history relates to the constellation of Cassiopeia, linking it with the name of Tycho Brahe, a noted Danish astronomer (1546-1601). In November 1572 an extremely brilliant star blazed forth in this region. The star became so bright that it was visible even in daytime. Brahe had fallen under the spell of astronomy early in life. He was so captivated by this unprecedented happening in a sky which was supposed to be unchanging that he began a systematic study of the star. Telescopes had not yet been invented, but Brahe carefully recorded light changes with the naked eye and made critical measurements of its position. For his work the star was named in his honor. Even though "Tycho's star" has now faded from view, the invaluable records which Brahe kept allowed modern astronomers to find the remnants of its invisible radio shock waves. These have been measured and are known to expand, even today, at the rate of 5,000 miles (8,000 kilometers) per second. Tycho's star was a supernova (page 120), the total explosive disintegration of a star, shining briefly with the light of 300 million suns.

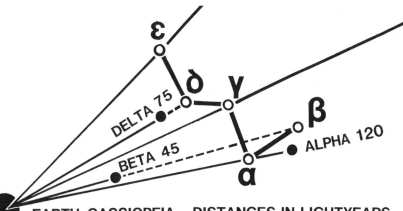

EARTH–CASSIOPEIA, DISTANCES IN LIGHTYEARS

GALACTIC CLUSTER

M 52

GALACTIC CLUSTER

M 103

107

PERSEUS PAGE 116

1h

Find the pictogram stars on this page and connect them with ruled colored lines. You can also create your own ideograms with red markings which will filter out and disappear under a red flashlight (page 135).

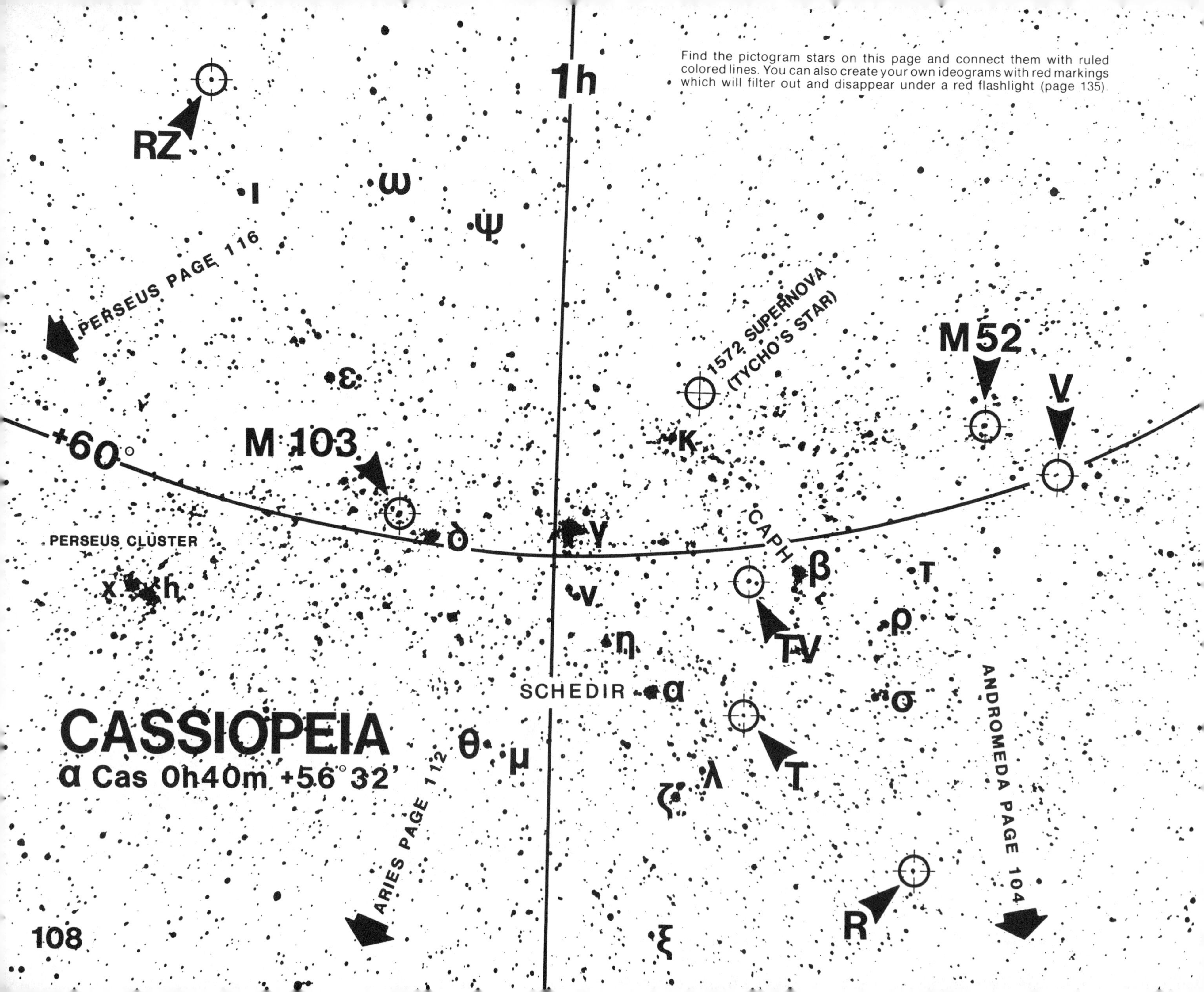

RZ

ι

ω

ψ

1572 SUPERNOVA
(TYCHO'S STAR)

M 52

V

ε

M 103

κ

+60°

δ

PERSEUS CLUSTER

γ

χ h

ν

CAPH

β

τ

ρ

η

TV

SCHEDIR

α

σ

CASSIOPEIA

θ

μ

λ

T

α Cas 0h40m +56°32'

ζ

ARIES PAGE 112

ANDROMEDA PAGE 104

R

ξ

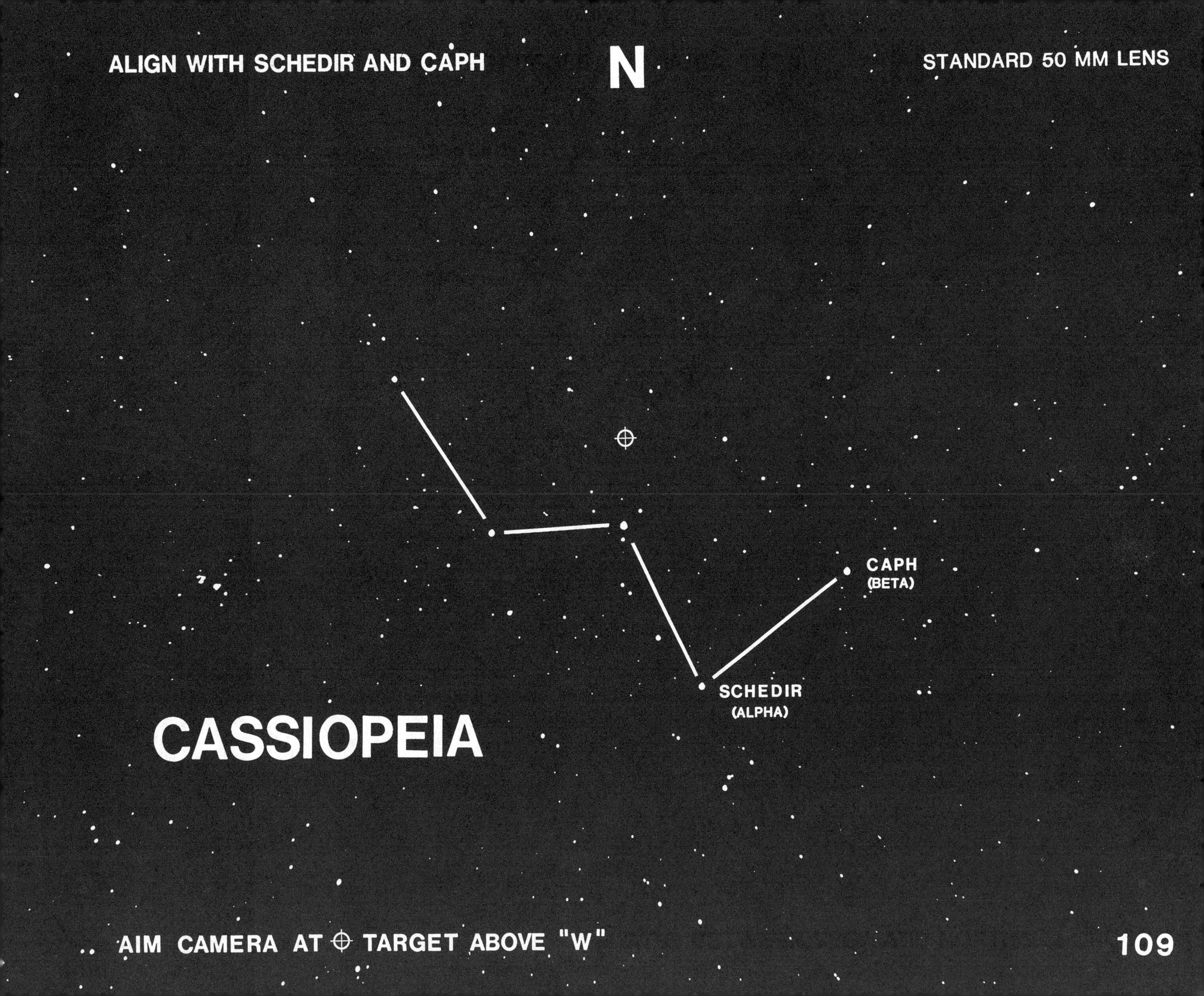

ALIGN WITH SCHEDIR AND CAPH

N

STANDARD 50 MM LENS

CAPH
(BETA)

SCHEDIR
(ALPHA)

CASSIOPEIA

AIM CAMERA AT ⊕ TARGET ABOVE "W"

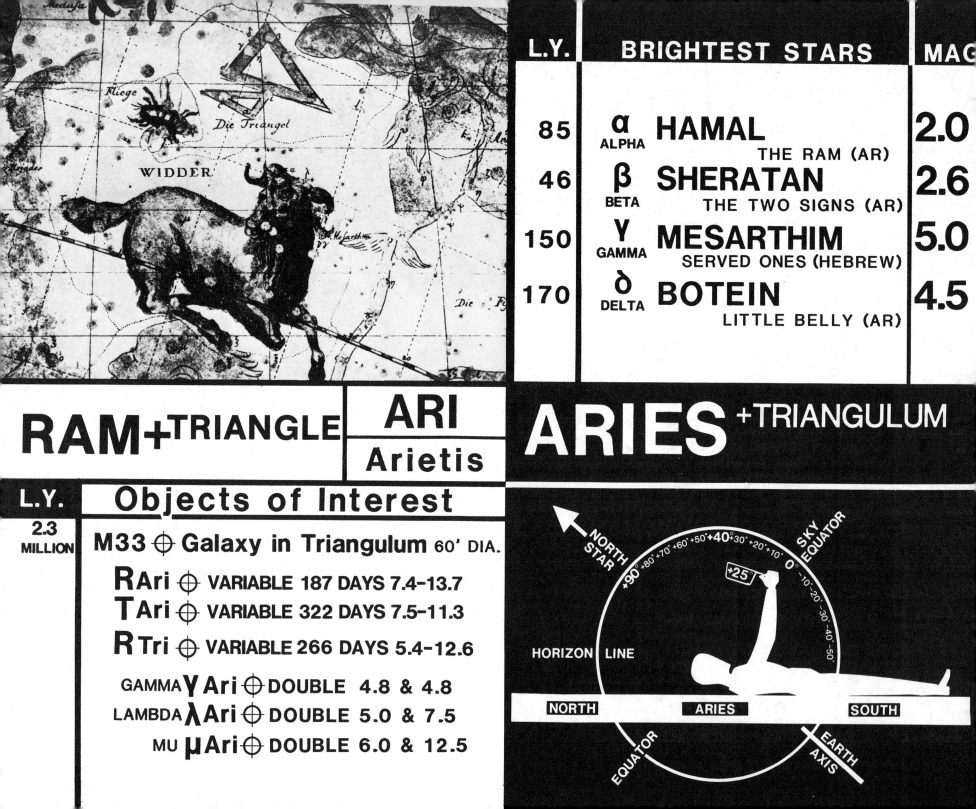

L.Y.	BRIGHTEST STARS	MAG
85	α ALPHA **HAMAL** THE RAM (AR)	2.0
46	β BETA **SHERATAN** THE TWO SIGNS (AR)	2.6
150	γ GAMMA **MESARTHIM** SERVED ONES (HEBREW)	5.0
170	δ DELTA **BOTEIN** LITTLE BELLY (AR)	4.5

RAM+TRIANGLE

ARI
Arietis

ARIES +TRIANGULUM

Objects of Interest

L.Y.	
2.3 MILLION	**M33** ⊕ **Galaxy in Triangulum** 60' DIA.
	R Ari ⊕ VARIABLE 187 DAYS 7.4-13.7
	T Ari ⊕ VARIABLE 322 DAYS 7.5-11.3
	R Tri ⊕ VARIABLE 266 DAYS 5.4-12.6
	GAMMA **γ Ari** ⊕ DOUBLE 4.8 & 4.8
	LAMBDA **λ Ari** ⊕ DOUBLE 5.0 & 7.5
	MU **μ Ari** ⊕ DOUBLE 6.0 & 12.5

NORTH STAR

+90° +80° +70° +60° +50° +40° +30° +20° +10° 0° -10° -20° -30° -40° -50°

SKY EQUATOR

+25°

HORIZON LINE

| NORTH | ARIES | SOUTH |

EQUATOR

EARTH AXIS

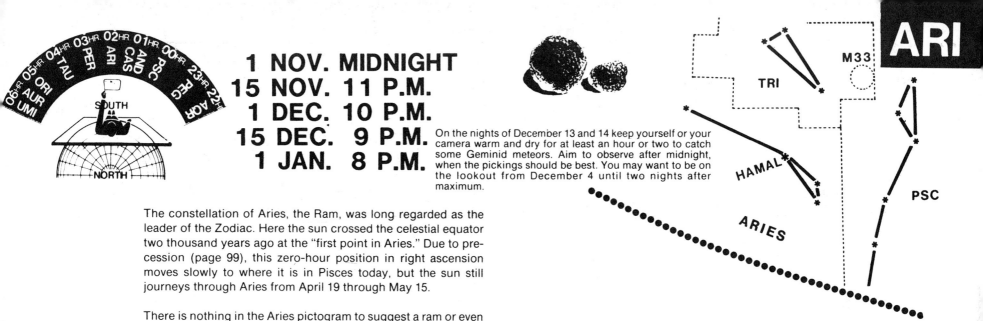

1 NOV. MIDNIGHT
15 NOV. 11 P.M.
1 DEC. 10 P.M.
15 DEC. 9 P.M.
1 JAN. 8 P.M.

On the nights of December 13 and 14 keep yourself or your camera warm and dry for at least an hour or two to catch some Geminid meteors. Aim to observe after midnight, when the pickings should be best. You may want to be on the lookout from December 4 until two nights after maximum.

The constellation of Aries, the Ram, was long regarded as the leader of the Zodiac. Here the sun crossed the celestial equator two thousand years ago at the "first point in Aries." Due to precession (page 99), this zero-hour position in right ascension moves slowly to where it is in Pisces today, but the sun still journeys through Aries from April 19 through May 15.

There is nothing in the Aries pictogram to suggest a ram or even an animal; however, the small constellation of Triangulum adjoining to the north allows immediate recognition and may be used to confirm the position of Alpha (α), Beta (β) and Gamma (γ) Arietis, which together also form a small flat triangle.

Mesarthim, Gamma (γ) Ari, is an "easy" double star with both components of 5th magnitude. Nearby Lambda (λ) Ari is a double of 5th- and 7th-magnitude stars with greater separation. The long-period variables R Ari and T Ari will reward the patient observer, as will R Trianguli.

There are no Messier objects in Aries, but M33, the famous spiral in Triangulum, is an important neighbor. This spiral galaxy, together with our Milky Way and with M31 in Andromeda, belongs to "our local group of galaxies" even though its distance is 2.3 million light-years. As with so many faint celestial objects, the low magnification of binoculars will be most useful in locating faint "island universes" such as M31 and M33 (page 118).

The remoteness of the brightest star in Aries was established through triangulation, whereby distance can be computed if the length of a baseline and two angles are known. This measuring method can be employed only for relatively nearby stars and uses the known diameter of the orbit of the earth around the sun as the basic measurement. That is how we learned that Alpha (α) Aries lies 75 light-years away from us.

The star Hamal will always bring back memories to me of the time I first viewed it. I had sought it out on the December evening of the day my mother died at the age of seventy-five. As I watched this distant sun blazing in the heavens, I became aware that the starlight I was seeing that solemn night had been on its way to my eyes — at a speed of 186,000 miles per second — for the entire time of my mother's life.

SPIRAL GALAXY

M33 GALAXY IN TRIANGULUM

THE SUN CROSSED THE CELESTIAL EQUATOR IN ARIES WHEN ASTROLOGY WAS FIRST INVENTED. TODAY THE EQUINOX LIES IN THE CONSTELLATION PISCES (PAGE 99).

111

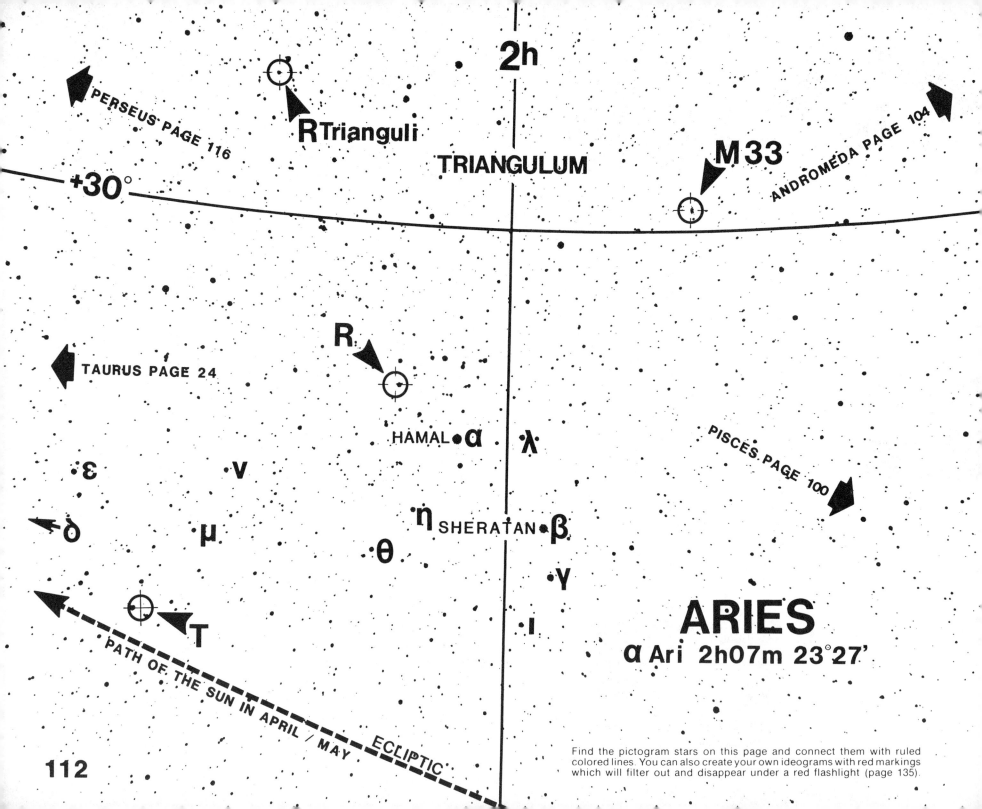

2h

R Trianguli

TRIANGULUM

M33

PERSEUS PAGE 116

ANDROMEDA PAGE 104

+30°

R

TAURUS PAGE 24

PISCES PAGE 100

HAMAL • α   λ

ε

ν

δ

μ

η SHERATAN • β

θ

γ

T

ι

ARIES
α Ari 2h07m 23°27'

PATH OF THE SUN IN APRIL / MAY   ECLIPTIC

Find the pictogram stars on this page and connect them with ruled colored lines. You can also create your own ideograms with red markings which will filter out and disappear under a red flashlight (page 135).

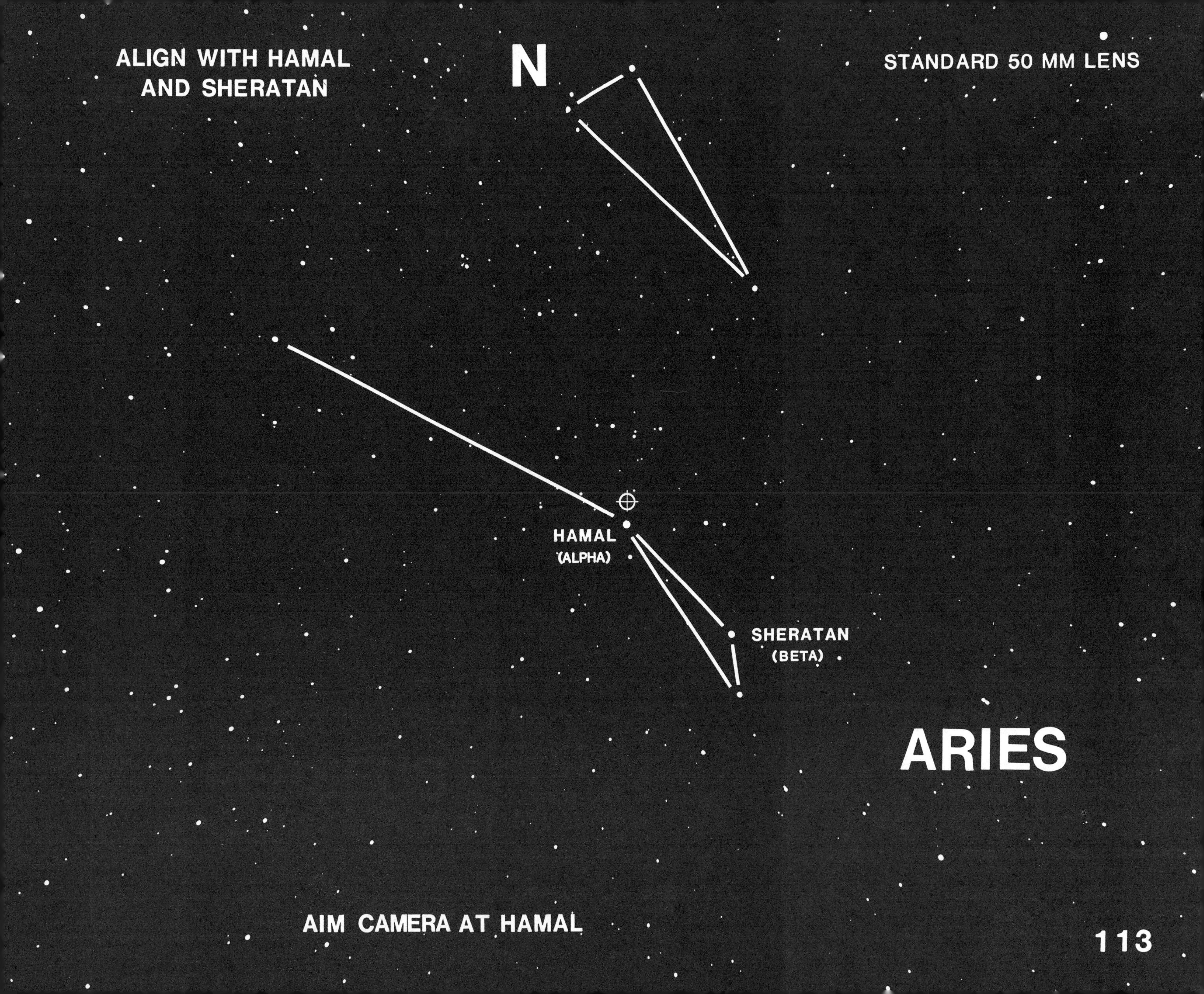

ALIGN WITH HAMAL
AND SHERATAN

N

STANDARD 50 MM LENS

HAMAL
(ALPHA)

SHERATAN
(BETA)

ARIES

AIM CAMERA AT HAMAL

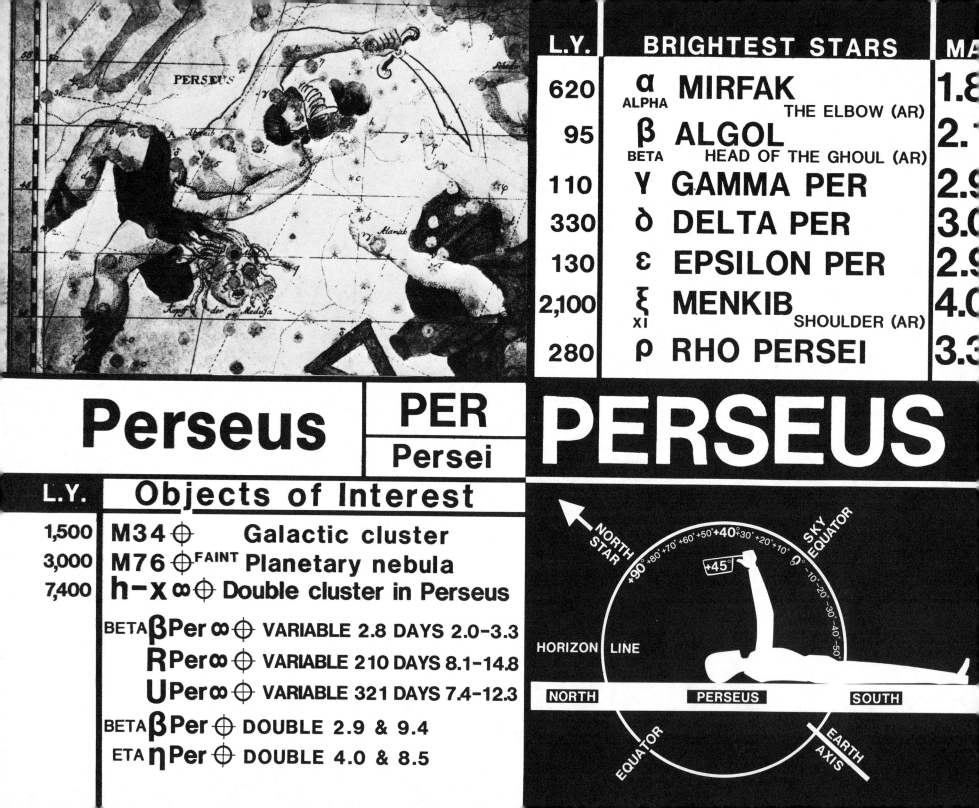

L.Y.	BRIGHTEST STARS		MA
620	α ALPHA	MIRFAK THE ELBOW (AR)	1.8
95	β BETA	ALGOL HEAD OF THE GHOUL (AR)	2.
110	γ	GAMMA PER	2.9
330	δ	DELTA PER	3.0
130	ε	EPSILON PER	2.9
2,100	ξ XI	MENKIB SHOULDER (AR)	4.0
280	ρ	RHO PERSEI	3.3

Perseus

PER / Persei

PERSEUS

L.Y.	Objects of Interest		
1,500	M34 ⊕	Galactic cluster	
3,000	M76 ⊕ FAINT	Planetary nebula	
7,400	h-χ ∞⊕	Double cluster in Perseus	
BETA	β Per ∞⊕	VARIABLE 2.8 DAYS 2.0-3.3	
	R Per ∞⊕	VARIABLE 210 DAYS 8.1-14.8	
	U Per ∞⊕	VARIABLE 321 DAYS 7.4-12.3	
BETA	β Per ⊕	DOUBLE 2.9 & 9.4	
ETA	η Per ⊕	DOUBLE 4.0 & 8.5	

NORTH STAR

SKY EQUATOR

+90° +80° +70° +60° +50° +40° +30° +20° +10° 0° -10° -20° -30° -40° -50°

+45°

HORIZON LINE

NORTH   PERSEUS   SOUTH

EQUATOR

EARTH AXIS

15 NOV. MIDNIGHT
1 DEC. 11 P.M.
15 DEC. 10 P.M.
1 JAN. 9 P.M.
15 JAN. 8 P.M.

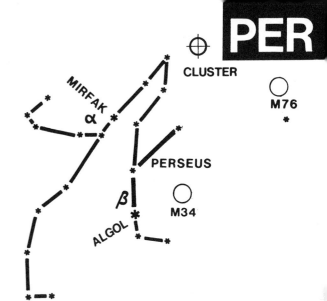

Make someone happy. Give a camera or a pair of binoculars for the holidays. They need not cost a fortune. Pawnshops often have quality items. Optics reasonably cared for cannot wear out. See pages 127-29 to find out why old-fashioned equipment will work just fine. Make yourself happy. Treat yourself to a simple camera: become a collector of starlight.

The constellation of Perseus is steeped in mythology. The legend of the hero surrounds this son of Zeus who slew the dangerous serpent-haired Medusa while he viewed her reflection in his polished shield to avoid the look that could kill. From his head, covered with a helmet of invisibility, to his toes, clad in winged sandals, Perseus was the complete champion. After beheading the hideous Gorgon, he went to rescue Andromeda, who, with Pegasus and Cassiopeia, was featured in the epic of this celestial conqueror. David and Golaith, even Saint George, slayer of the dragon, have been identified with the constellation.

By stretching the imagination in keeping with feats of our hero, we may be able to recognize a figure in the pictogram. Alpha (**α**) Persei is Mirfak, the elbow of Perseus. You may have to guess the rest. We need Bode's more literal interpretation to put Beta (**β**) Per into perspective, and to see the head of the ghastly Medusa whose eye could literally turn men to stone. The name "Algol" dates back to the Arabic *Al Ras al Ghul,* which in turn comes from the Greek of Ptolemy "the Head of the Ghoul." The sinister connotation suggests that already in the dark past the strange behav-

ior of Beta (**β**) Per had been noticed. The regular changes in magnitude which make the star appear to "blink" are visible even to the naked eye. The regularity of the variations in brightness were first recorded soon after the invention of the telescope. In the sixteenth century the theory was put forward that the dimming of Algol was caused by a dark companion revolving around the star and eclipsing it. This was confirmed scientifically just one hundred years ago, and today we know that Beta (**β**) Per, normally at magnitude 2.1, fades to magnitude 3.4 every 68 hours 48 minutes 56 seconds before gradually brightening again. It is an "eclipsing binary," and the eclipse lasts ten hours.

Messier lists his discovery of M34, a bright galactic cluster, and the much fainter planetary nebula M76 (page 79). For unexplained reasons, he makes no mention of the splendid "double star cluster in Perseus," easily within reach of his telescopes and our binoculars.

GALACTIC CLUSTER

M 34

PLANETARY NEBULA

M 76

ORBIT

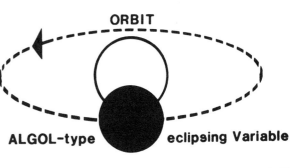

ALGOL-type   **eclipsing Variable**

SEE PAGE 121

115

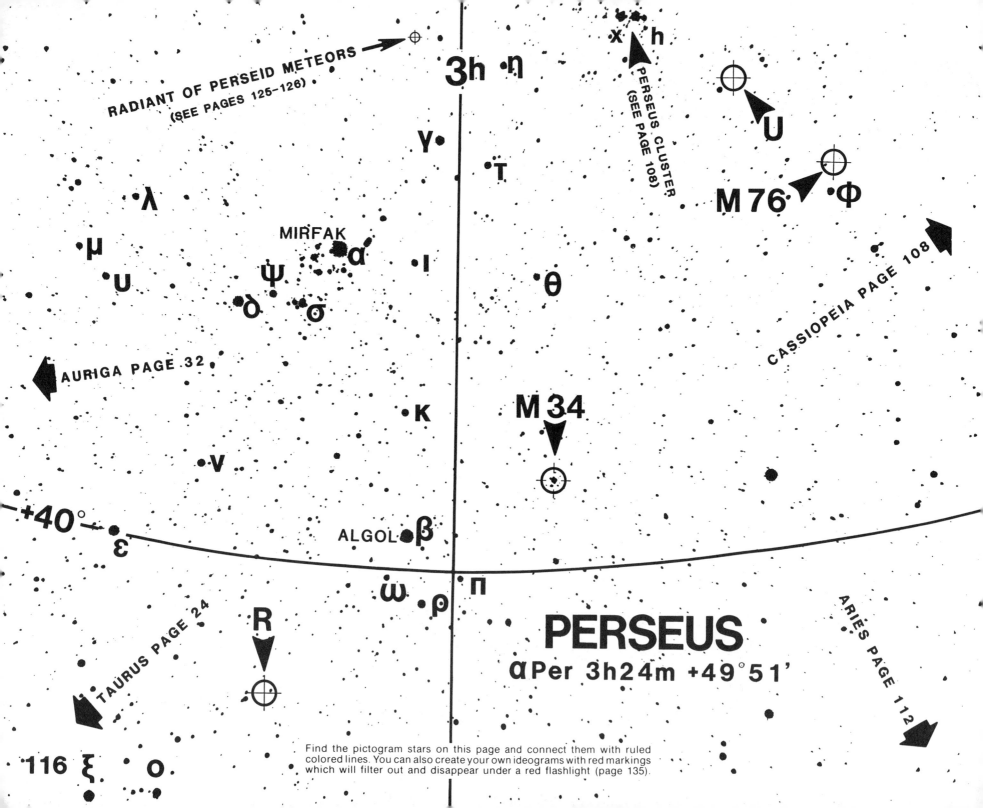

RADIANT OF PERSEID METEORS
(SEE PAGES 125-126)

3h •η

x h
PERSEUS CLUSTER
(SEE PAGE 108)

U

M 76
Φ

γ

•T

λ

μ

U

MIRFAK

α

ψ

•I

δ

σ

θ

CASSIOPEIA PAGE 108

AURIGA PAGE 32

K

M 34

V

+40°
ε

ALGOL
β

ω
ρ
Π

PERSEUS
αPer 3h24m +49°51'

ARIES PAGE 112

TAURUS PAGE 24

R

116 ξ
o

Find the pictogram stars on this page and connect them with ruled
colored lines. You can also create your own ideograms with red markings
which will filter out and disappear under a red flashlight (page 135).

ALIGN WITH MIRFAK AND ALGOL

MIRFAK
(ALPHA)

ALGOL
(BETA)

PERSEUS

GALAXIES
ISLAND UNIVERSES

Vast congregations of countless stars, nebulae, clusters and interstellar matter, are the stuff of which galaxies are made. Among the teeming multitudes of such assemblages of millions or hundred-thousands of millions of stars, our own galaxy is just one. Even with the finest and latest equipment we can only dimly perceive some of the profusion of galaxies which abide in the universe. Hosts of great island universes lie far beyond the reach of our feeble equipment and faculties.

It is most fitting that the term "island universes" was first coined to describe galaxies by a philosopher rather than an astronomer, by a genius who viewed the heavens with his mind's eye instead of a telescope. Immanuel Kant never traveled farther than 60 miles from the Prussian town where he was born in 1724. Kant's vision extended beyond the obvious. His original thinking can be compared to that of early Greeks like Plato and Aristotle. These were men whose perceptions reached far beyond their own place and time.

Position of
OUR SOLAR SYSTEM
(PAGES 14–15)

GLOBULAR CLUSTERS

100,000,000 +

LIGHT YEARS

OUR GALACTIC CENTER

100,000 LIGHT YEARS

2,200,000 LIGHT YEARS

OUR GALAXY
(THE MILKY WAY)

GALACTIC (OPEN) CLUSTERS

M31

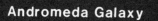
Andromeda Galaxy

Spiral   **Elliptical**

Barred spiral

As recently as one hundred years ago the hazy patch in Andromeda to which Messier assigned the number 31 was thought to be a close nebula, a rare cloud of gas without much substance. It was the work of the American Edwin Hubble which allowed us to better set our sights. By studying certain variable stars (page 121) which reveal their distance through periodic changes in brightness, Hubble was able to establish, in 1925, that what had seemed like a tenuous nearby nebula was in fact a gigantic galaxy of suns far, far beyond our own galaxy. Today we know that the distance to the Andromeda spiral is on the order of 2.2 million light years. Even at this mind-boggling range M31 is regarded as a member of our "local group" of galaxies, to which M33, the spiral in Triangulum (page 112) and our own Milky Way all belong.

In time this research led to a discovery which will bear the name "Hubble's Law" forever. It establishes that the universe is expanding. Even though we are nowhere near the center of this galactic bubble, all galaxies keep moving away from one another and from us at a speed which is proportional to their distance. The farther out they are, the faster they zoom away, like polka dots printed on a gradually inflating balloon. Cosmologists now debate whether this expansion will continue indefinitely or whether it will one day slow down and reverse itself.

Self-evident truths, called axioms, permit us to determine distances to galaxies even further out. Axioms are based on earlier learning and on deductive reasoning. Thus, if A equals B and B equals C, then A equals C. In astronomy, logic is interwoven with careful observations to increase our knowledge.

Galaxies come in many different shapes and sizes. There are spirals, barred spirals and ellipticals. Some are seen edge-on, like Frisbees in flight, others resemble revolving pinwheels.

When you consider the immense astronomical cosmos where frontiers are daily set farther and farther out, it seems only natural to contemplate the microcosm at the opposite end of the enormous scale. Here too endless space lies far beyond our perceptions.

As we soar among the galaxies let us sometimes meditate on these other worlds within — worlds without end.

NOVAE AND SUPERNOVAE

A nova is a star which suddenly erupts and emits a great outburst of light. Nova literally means new. Stars which undergo such changes could not be observed in ancient times until they exploded and could be seen by the naked eye. This made them appear as new stars. The explosive violence of novae (plural of nova) increases the luminosity of such stars by up to tens of thousands of times. Nova spectaculars are dwarfed by supernovae where stars undergo flare-ups of up to hundreds of millions of times in brightness.

Records tell us that novae occur in our galaxy at the rate of a dozen or so per year. There may be many more. Most escape detection because few stargazers are as yet patrolling wide starfields with cameras and the simple new Problicom (pages 132-33) to discover them. Professionals work only on specific projects in minute areas and lack the time for important but unpredictable searches. Their huge narrow-field telescopes are useless for broad surveys, but once an amateur has made a discovery, astronomers all over the world swing their mammoth mirrors to wrench secrets from their new common target. Such research now suggests that novae always involve a pair of stars very close together which interact until one of them blasts away its outer layers in a titanic explosion of stunning brilliance which may remain visible for a few days or weeks. Novae which recur bring them into the realm of variable stars (page 121).

The Crab Nebula in Taurus (page 23) and Tycho's star in Cassiopeia (page 107) bear witness to two of the three most famous supernovae which have been reported in our galaxy in the last ten centuries. What supernovae lack in frequency (about one in three hundred years) they make up in the violence which marks the collapse and near disintegration of an aging star. It would not be surprising if our own galaxy or one of our neighboring galaxies produced such a spectacular phenomenon at almost any time.

We have always been fascinated or awed by grand displays of power: the deafening crack of thunder caused by lightning, the earth-shaking roar of engines lifting a spacecraft from the pad and lofting it into orbit, the silent, constant and limitless outpouring of energy from our sun. When observing a nova or a supernova, we are witnessing the ultimate release of unbridled natural force. In a few dozen hours the amount of energy which our searing sun produces over a period of millions of years is triggered in an unimaginable outburst. The doomed star will shine with dazzling brightness, night and day, before its last glorious fire is damped forever in the absolute cold of space.

Even in death such a star will have formed an ever-growing new shell of gas filled with elements. the building blocks for new beginnings.

SPIRAL GALAXY M 100

NOVA CYGNI 1975 NEAR MAXIMUM

NOVA FADING 6 MONTHS LATER

M 100 WITH SUPERNOVA

VARIABLE STARS

The sky is filled with many surprises. Constant change is the order of the night. The dazzling light of a nova which occurred on the day you were born may even now be racing through interstellar space at 186,000 miles per second to arrive here next Friday and present to you an opportunity for discovery. Thousands of variable stars are at this moment going through titanic upheavals in the seething turbulence of their ever-changing nuclear fires.

Most stars shine with constant brightness, their magnitudes fixed eons ago. But some stars, known as variables, change in luminosity over periods of time which range from hours to hundreds of days. Many of the brighter variables are easy to monitor and allow us to wrest important secrets from the universe. That is why amateur astronomers have begun to play an increasing part in this exciting field.

Observations build on the far-reaching discovery made by the American astronomer Henrietta Leavitt. In 1912 she found that the changing brightness of certain variables, when related to their period (the time such stars take to go from bright to faint and back again), allows us to determine star distances.

Although many variable stars are well recorded, there are great numbers of so called "irregular variables," as well as others which need constant monitoring to learn more about their behavior and special characteristics. Here is an area where dedicated stargazers can make truly valuable contributions to science.

The best and easiest way to find variables is to start photographing a starfield (centering the camera on the same bright star) at one-month intervals. Exposures should always be of the same lengths. Please take notes (see pages 127-31). Then visually compare photos where bright variables are indicated. For best results use a Problicom to "blink" your slides. You are bound to find many more variables than are listed under "objects of interest."

The American Association of Variable Star Observers (A.A.V.S.O.) is a scientific and educational nonprofit organization founded in 1911 to promote and coordinate such observations. Anyone is welcome to join this group of starlovers. The data which individuals collect and record are combined by the A.A.V.S.O. with other observations. Then light-curves are prepared for distribution to the astronomical community. In this manner, hundreds of thousands of separate observations with the naked eye, binoculars or telescopes together become important components in the increase of human knowledge about stars, our galaxy and the universe at large. Your own conscientious observations and reports are advanced almost instantly to the very frontiers of science. The world's leading astronomers often call on the A.A.V.S.O. to have members monitor specific variables. This can be in connection with a planned space launch or some other new research project.

If you are interested in exploring the universe and learning more about the A.A.V.S.O., send a self-addressed 9½-by-4⅛-inch double-stamped envelope to the American Association of Variable Star Observers, 187 Concord Avenue, Cambridge, MA 02138. In due course you will receive information on how you can participate in important and useful astronomical science projects.

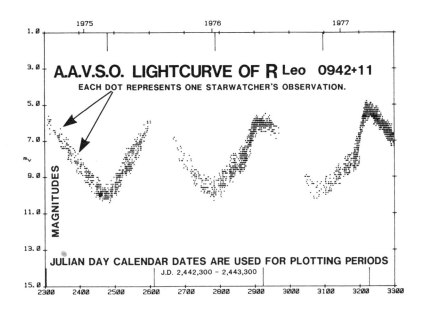

A.A.V.S.O. LIGHTCURVE OF R Leo 0942+11
EACH DOT REPRESENTS ONE STARWATCHER'S OBSERVATION.

JULIAN DAY CALENDAR DATES ARE USED FOR PLOTTING PERIODS
J.D. 2,442,300 - 2,443,300

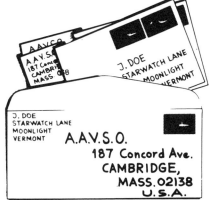

J. DOE
STARWATCH LANE
MOONLIGHT
VERMONT
A.A.V.S.O.
187 Concord Ave.
CAMBRIDGE,
MASS. 02138
U.S.A.

A.A.V.S.O. CHARTS

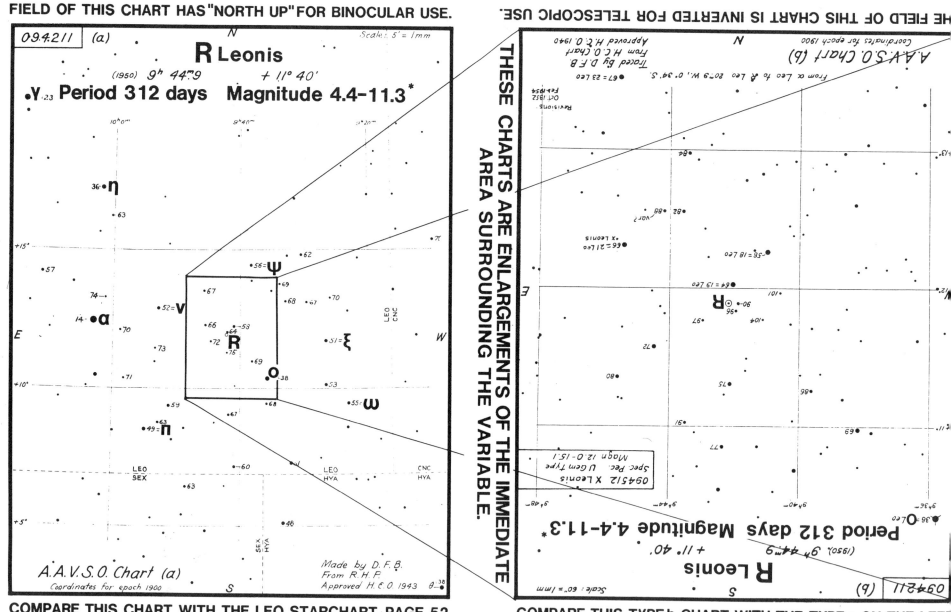

THESE CHARTS ARE ENLARGEMENTS OF THE IMMEDIATE AREA SURROUNDING THE VARIABLE.

COMPARE THIS CHART WITH THE LEO STARCHART, PAGE 52.

Numbers beside various stars are the magnitudes of stars that may be used for comparison with the variable. The decimal point is omitted, it might be mistaken for a star. Thus, **64** indicates magnitude **6.4**, **101** signifies **10.1**, and so on.

COMPARE THIS TYPE b CHART WITH THE TYPE a ON THE LEFT

On the AAVSO charts magnitudes are indicated to tenths. By comparing the variable to comparison stars, accurate brightness estimates can be made. Do not forget, the larger the number, the fainter the star. *1976 KUKARKIN VARIABLE STAR CATALOGUE

EUTERPE ▶

ASTEROIDS

Asteroids are also called "minor planets." In the planet diagram (pages 14-15) a broad band is shown indicating the space between the orbits of Mars and Jupiter where many of these miniplanets travel. It is called the "asteroid belt." Ceres (#1), the largest of these planet chunks, was the first to be discovered, in the year 1801. It is estimated to be about the size of the state of Michigan in diameter and roughly spherical in shape. Pallas (#2), Juno (#3) and Vesta (#4) were next discovered. Astronomers have studied and computed the orbits of about 2,200 of these planetoids and assigned them numbers in the order of their discovery. At first, classical Greek names were chosen for them, but eventually modern designations were used to honor famous persons, events or even places. There are asteroids named Mozartia (#1034) and Tombaugh* (#1604), Armisticia (#1464) and Olympiada (#1022), as well as Pasadena (#2200) and Tucson (#2224). It is estimated that there are about 100,000 minor planets coasting around the sun. Most are only one mile or a kilometer across, but, because many are thought to be abundant in metals, space experts are exploring the possibility of capturing asteroids so that they can be mined for precious ores.

Just like the major planets or their satellites, asteroids merely reflect the light of our sun. For this reason brightness can vary depending on the asteroids' size, their reflectivity and their distance. Some minor planets have orbits which bring them very close to earth, putting them on the borderline of naked-eye visibility. Astronomers in the Soviet Union assist the international astronomical community by each year publishing exact tables showing the positions, at regular intervals, of these odd-shaped celestial Gibraltars.

During photographic sky surveys or nova patrols, asteroids will often be recorded as shown in the pair of photographs below. Their motions make them jump back and forth for immediate detection when such a twenty-four-hour twosome is "blinked" in a Problicom (pages 132-33). In long exposures through telescopes, motions will leave trails of light. The asteroid Euterpe (#27) was accidentally recorded just south of the Crab nebula M1, during a one-hour time exposure as shown.

*The American astronomer Clyde Tombaugh discovered the faint ninth planet, named Pluto, near Delta (δ) Geminorium (page 40). He made his discovery during the course of a systematic planet search by blinking photographs taken on different nights. Tombaugh's place in the history of astronomy is assured for his diligence and perseverance in a quest which took years to complete.

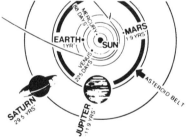

ENLARGEMENTS FROM 15 MIN. EXPOSURE COLORSLIDES TAKEN 24 HOURS APART.

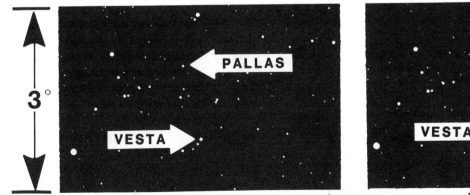

BRIGHTEST STAR IN FIELD IS FIFTH MAGNITUDE OMEGA¹ (ω^1) AQUARII.

ИНСТИТУТ ТЕОРЕТИЧЕСКОЙ АСТРОНОМИИ
АКАДЕМИИ НАУК СССР

ЭФЕМЕРИДЫ МАЛЫХ ПЛАНЕТ

на 1982 год

EPHEMERIDES OF MINOR PLANETS

for 1982

EDMUND HALLEY
1656 – 1742

COMETS

Comets are the stuff from which astronomical dreams of fame are made. The finding of a comet automatically bestows the ultimate honor on the discoverer, whose name will be joined with the cosmic snowball for all time. Here is an arena in which many amateurs have forever linked their signature to the skies.

Dusk and dawn are the times when dedicated comet seekers look for blurry objects in what have come to be known as the "haystacks" (pages 140-41), although you may also trap a comet at the zenith some midnight while searching for novae. A "haystack" is an area approximately as outlined on page 141, where an evening comet-search area is shown, and on page 140, which delineates a morning haystack profile. Comets are usually first discovered going into their turn around the sun or coming out of it. That is what suggests times and places for best results. It is no longer necessary for the comet hunter systematically to scan back and forth for an hour or two while gradually moving binoculars or telescope up from the horizon (or down to it). A few carefully timed and framed photographs can accumulate the elusive light of a fuzzy comet in its early stages or the plume of its tail, always pointing away from the sun. If you shoot with fast Tri-X black-and-white film you can develop the negatives and blink-compare them immediately in a home-built Problicom (pages 132-33). This will make you a serious contender in the comet marathon, where most entrants still work by visual observation alone (you can combine both).

Edmund Halley, the Astronomer Royal at the famed Greenwich Observatory in England, was the first to figure out that comets move in cigar-shaped orbits and that some may keep returning to swing around the sun periodically (page 14). In the year 1695 he used Isaac Newton's new mathematics to compute that the "hairy stars" seen in 1531, 1607 and 1682 were one and the same, appearing about every seventy-six years. He correctly predicted — but did not live to see — the return of the roamer in 1758, when it was named "Halley's Comet" in his honor. It is the best known of the "periodic" comets. Almanacs list their comings and goings.

Other comets take just one turn around the sun before heading back into deep space, not to return for thousands or millions of years. Astronomical publications report their changing whereabouts. Even if you cannot find one at first because of its low brightness, time exposures in the general direction of its host constellation can reveal the position of fuzzy nuclei to the 10th magnitude.

Astronomers believe that the heads of comets consist of ices mixed with rocks and dust which date back to the formation of our solar system. Every time a nucleus moves close to the sun some of its mass is pushed into a tail of dust and gas by the pressure of sunlight.

ROYAL OBSERVATORY
GREENWICH, ENGLAND.

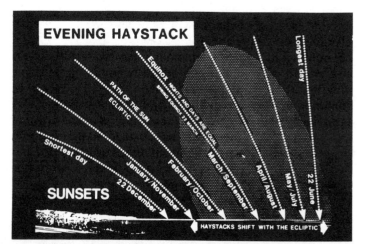

EVENING HAYSTACK

Longest day
Equinox, nights and days are equal
SPRING EQUINOX, 21 MARCH
PATH OF THE SUN
ECLIPTIC
Shortest day
January/November
22 December
February/October
March/September
April/August
May/July
22 June
SUNSETS
HAYSTACKS SHIFT WITH THE ECLIPTIC

METEORS

SEE PAGE 52

LEONID RADIANT

There is a relationship between the slow-moving glow of distant comets and the swift needle-sharp brilliance of Meteors, also called shooting stars. Mixed within the frosty mantle which may surround the core of large iceberg-sized comets are sand-grain-sized granules of rock. These leave invisible trails of debris in the comet's wake. Many, drawn earthward by gravity, enter our atmosphere from all directions. These fast-moving particles are heated by friction to where they briefly flare with incandescence before vaporizing at heights ranging from 100 to 150 miles. Such random meteors by themselves are called "sporadics." Larger pieces may make it down to the ground, in which case they are called "meteorites."

At certain times of the year, the earth in its journey around the sun coasts through the debris-littered orbit of a comet as shown on the next page. On such occasions we encounter "Meteor showers," where dozens or even hundreds of shooting stars are visible in one hour. They give the appearance that they all originate from one point in the constellation after which the shower is named. This source, called the "radiant," is the result of an optical illusion much like the vanishing point of roads or telegraph poles coming from a great distance.

Photography of meteors is easy (pages 126-27) and can yield spectacular results. You will see the greatest number of shooting stars after midnight, when the earth will have turned to where it "faces the meteor

REGULUS

METEORITES (IRON)

CENTIMETERS

traffic." It can be compared to a car heading into the rain, with many more raindrops hitting the front windshield than the rear. Meteors shower down on earth at the rate of some 300 tons per day. Many fall into the oceans, and most are small. Some meteors are pieces of asteroids. Occasionally a major meteorite impact is reported in the media. The best known in the United States may be the meteor crater near Flagstaff, Arizona, which dates back to prehistoric times.

GEMINID ⊕ RADIANT

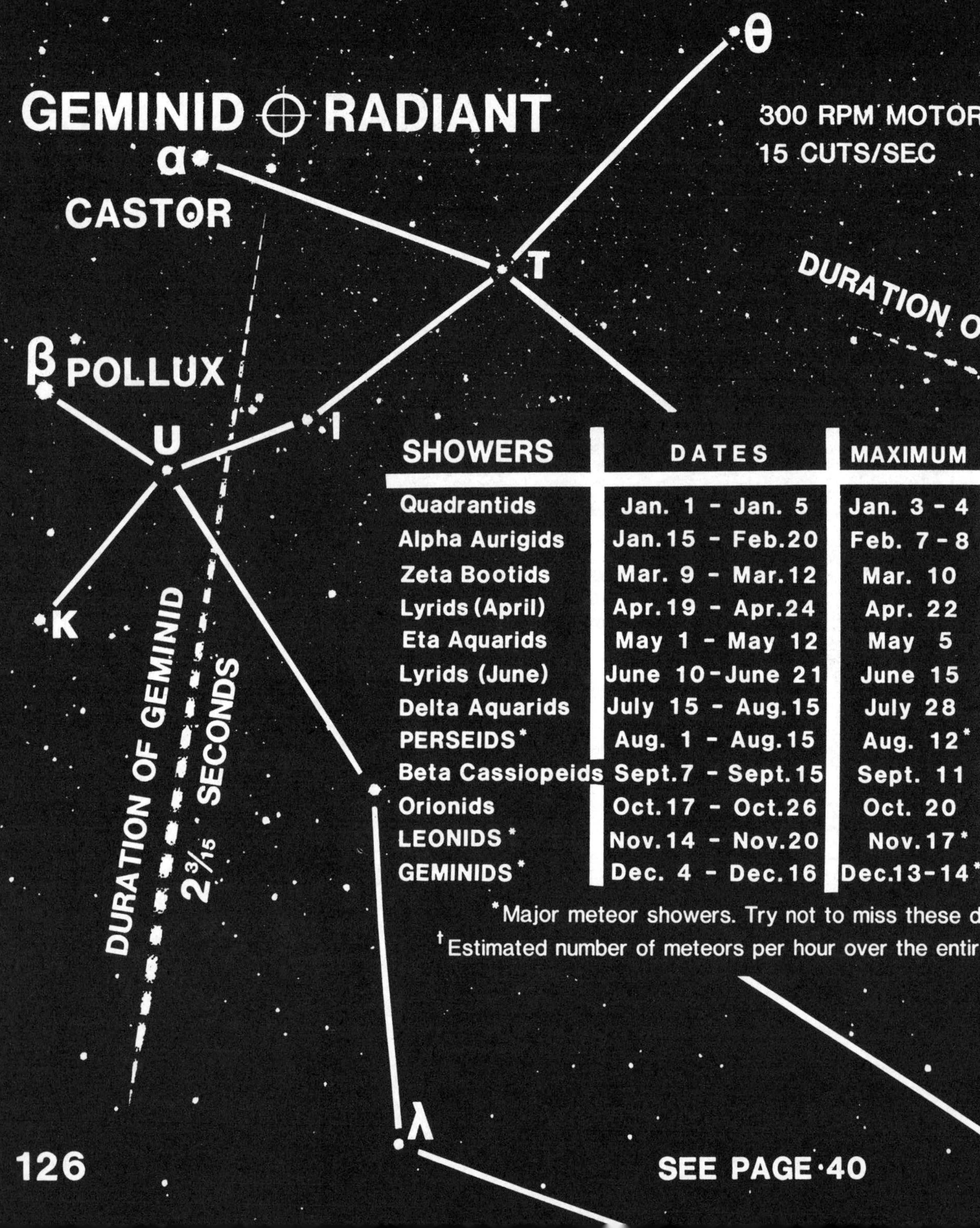

α CASTOR

β POLLUX

θ

T

U

I

K

λ

γ

μ **η**

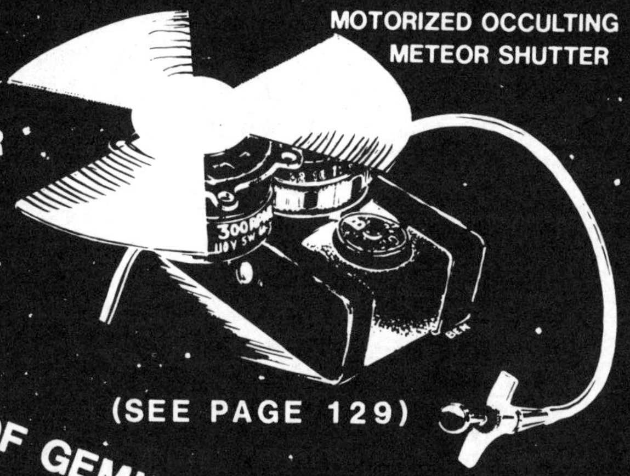

MOTORIZED OCCULTING METEOR SHUTTER

300 RPM MOTOR
15 CUTS/SEC

(SEE PAGE 129)

DURATION OF GEMINID $1\frac{3}{15}$ SECONDS

DURATION OF GEMINID $2\frac{3}{15}$ SECONDS

SHOWERS	DATES	MAXIMUM	METEORS[†]
Quadrantids	Jan. 1 - Jan. 5	Jan. 3 - 4	20 - 80
Alpha Aurigids	Jan.15 - Feb.20	Feb. 7 - 8	12
Zeta Bootids	Mar. 9 - Mar.12	Mar. 10	10
Lyrids (April)	Apr.19 - Apr.24	Apr. 22	12
Eta Aquarids	May 1 - May 12	May 5	20
Lyrids (June)	June 10 - June 21	June 15	15
Delta Aquarids	July 15 - Aug.15	July 28	35
PERSEIDS*	Aug. 1 - Aug. 15	Aug. 12*	65
Beta Cassiopeids	Sept.7 - Sept.15	Sept. 11	10
Orionids	Oct.17 - Oct.26	Oct. 20	35
LEONIDS*	Nov.14 - Nov.20	Nov.17*	10-100
GEMINIDS*	Dec. 4 - Dec.16	Dec.13-14*	50

*Major meteor showers. Try not to miss these dates!
[†] Estimated number of meteors per hour over the entire sky.

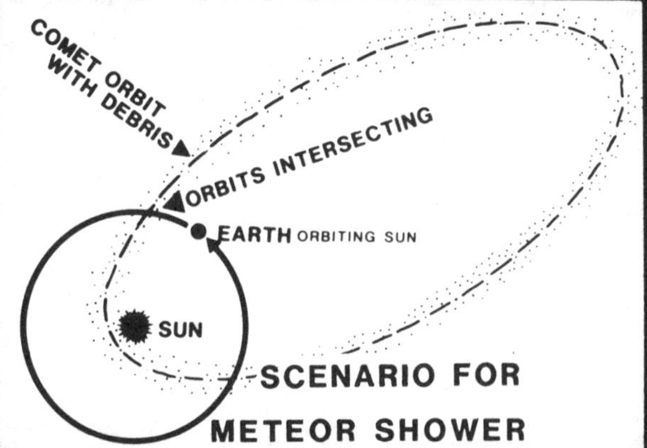

SEE PAGE·40

COMET ORBIT WITH DEBRIS ▶

ORBITS INTERSECTING

EARTH ORBITING SUN

SUN

SCENARIO FOR METEOR SHOWER

SIMPLE ASTROPHOTOGRAPHY

You too can collect starlight.

Basic astrophotography is not difficult. It will allow anyone to perform important work, even to make discoveries, with simple equipment. Have a loaded camera with the shutter open pointed at the night sky as often as possible. Photographic lenses in leather cases, just like lures in a tackle box, will never make catches. Cameras must be ready and pointed skyward before a prize can be captured.

One might think that expensive camera equipment is needed and that only the latest would be of use. Quite the opposite is the case. Cameras should accept 35 mm. films. The shutter must have a B (bulb) setting so that it can be held open with a cable release for varying periods of time and then closed again. The lens can be any standard type, which means that focal length would be around 50 millimeters. (Lens information is always prominently engraved on the front rim). Finally, you should get a clear view of a distant horizon when you look through the lens via the viewfinder.

Most old or secondhand cameras perform fine service as long as the above requirements can be met. Consider buying a used camera and lens in a photo-supply store or a pawnshop. You will quickly see for yourself that there is one factor which will determine the cost of a good lens; it's speed. In astrophotography speed is the name of the game. It simply means that great amounts of starlight can pass through glass and widest aperture in the shortest period of time. It is usually expressed as "f" speed. For simplicity, think of "f" as meaning fastness.

Film Speed: LOAD FASTEST SLIDE FILM

* PRINT FROM A COLORSLIDE.
   FILM: HIGH-SPEED EKTACHROME ISO 400.
   EXPOSURE: 30 SECONDS.
   APERTURE: F 1.4, STANDARD LENS.
   RECORDS STARS TO SIXTH MAGNITUDE+.

BIG DIPPER PHOTOGRAPH* FROM TRIPOD.

HOLD CAMERA STEADY WITH A TRIPOD OR **Camera Clamp**

CABLE LOCK & RELEASE

127

Focusing Ring.

The numbers 1.4/50 engraved on the front ring of a lens mark a very fast 50mm lens. The widest opening on the aperture ring is then also marked 1.4. A 1.8/50 is a fast standard lens also, as is a 2/50.

All starframe photographs in this album were taken with a simple SLR (single lens reflex) camera using a standard 1.4/50mm lens. To capture larger fields, a slightly slower wide angle 2/35mm lens was used where noted. On page 56 the differences in the sizes of fields are shown.

Weather and moon permitting, you can begin tonight to take starfield photographs which will show the main constellations —even from a city. Here is what you do: Ask for and load the fastest color film (ISO 1000 is excellent). Color-transparency slide film offers the greatest versatility for our purpose. It can be turned into prints like the starfield photos in this book. When projected in a dark room, slides permit huge enlargement with planetarium fidelity. Slides also admit you to the astonishing world of Probli-com blinking (pages 132-33) where a genuine discovery may await you in every 2-by-2-inch cardboard frame.

To begin, point the lens at the sky on the dates and at the times as listed on pages 22-117. Secure the camera firmly after aiming it in the direction of the declination angle as illustrated for the star-frames. First expose two blank frames while briefly shining your flashlight near the lens to mark the beginning of your sequence and to indicate the spacing of your night photographs for the processing lab.

Focusing presents no problems: *Always* set your lens at infinity.

Aperture (lens opening) is easy: *Always* use your lens at its widest possible setting, the f/1.4, 1.8, or 2 opening.

Exposure is the only variable with which you will experiment. It offers many choices. Try this method for a start:

Focusing Ring: ALWAYS AT INFINITY

Aperture Ring: ALWAYS WIDE OPEN

Exposure Dial: ALWAYS AT B

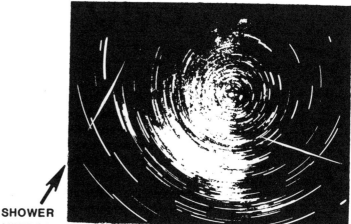

NORTH STAR REGION PHOTOGRAPH, WITH CAMERA CLAMP AND LADDER. FILM: KODAK TRI-X PAN, ISO 400. EXPOSURE: 60 MINUTES. APERTURE: F 2.2, STANDARD LENS. GOOD TARGET FOR METEOR SHOWERS.

SHOWER METEOR

SPORADIC METEOR

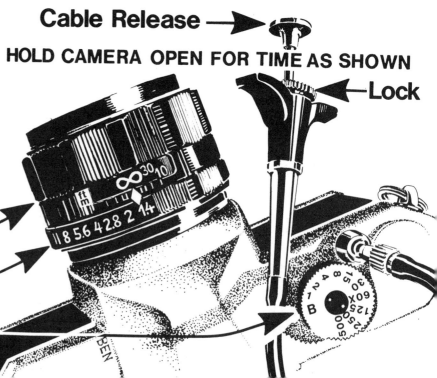

Cable Release ⟶

HOLD CAMERA OPEN FOR TIME AS SHOWN

⟵ **Lock**

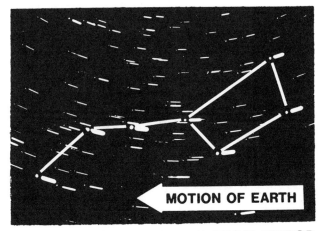

BIG DIPPER PHOTOGRAPH FROM TRIPOD.
FILM: FUJICHROME ISO 400
EXPOSURE: 20 SECONDS + 8 MINUTES.
APERTURE: WIDE OPEN, STANDARD LENS.
SEE TEXT FOR DETAILS ON THIS SLIDE.

Make your first exposure one second long. Double the next. Redouble and shoot for four seconds. Next expose for eight, then fifteen seconds. Now try a half-minute exposure and finally take one a full minute long. Whatever you do, keep an exact record of every photograph in a notebook. You will find that, as the exposures become longer, more and more fainter stars register, until eventually round star images turn into little star trails. When you aim your camera at the North Star, 15, 30, or 60 *minute* exposures will pinpoint the position of the North Celestial Pole (N.C.P.). Such longer exposures are what you use during meteor showers to record shooting stars. They will show star trails and underline the differences in star colors.

To detect the direction of the rotation of the earth, try the following: With your cable release, lock the camera open for an exposure of about 20 seconds. Do not close the camera yet, but block any more light from entering it by holding a piece of cardboard over the lens for one minute. Do not touch the camera. Next remove the occulting piece of card and continue with the exposure for an additional 8 minutes. Then allow the shutter to close. The resulting photograph will establish which way the earth is turning because you can see the starting star points on your photograph. A motorized rotating occulting shutter in front of an open camera lens (page 126) will chop fast meteor trails into interesting mini-exposures.

You may wish to watch for meteors while photographing. After a while you may want to wrap your camera in a plastic bag, cutting openings for the lens and cable release, to protect your system from humidity. Then you can open your shutter and go back indoors while your starlight collector keeps a meteor vigil for you. Set an egg timer to remind you to go out every 10, 20 or 30 minutes to close the shutter, advance the film and reopen the camera. Check for dew on the lens. A hair drier aimed at an angle toward your camera will keep the optics good and dry.

You may not know it until later, but in any of your exposures you may have accidentally recorded an important celestial event. It happened to me (see Nova Cygni, page 83). This is only one of the reasons why you should keep accurate records from the very beginning. The Smithsonian asked me for the date and the hour when I took each photo, as well as the exposure times and the type of film, camera and lens. Your own written notes will become the most valuable manual for all your future photography. You will not be able to remember tomorrow just when and for how long you exposed your pictures last night. Experimenting with new ideas will allow you to see what a change in exposure time actually did to your film. You will need the notes when you hold the developed images in your hand. The best way to learn is by trial and error. Not a single frame will be wasted, especially when you start comparing slides of the same region (of similar exposure times) which you will have taken on different dates. Such comparison is easy with a home-built discovery machine (pages 132-33).

Date and number each roll of film. Write your constellation target on each slide. Indicate which way is north. Keep and file every slide. A seemingly uninteresting photo taken six weeks ago may hold a hidden treasure. A fine meteor photograph will be its own reward.

IMPORTANT: ALWAYS KEEP
DETAILED RECORDS OF
EVERY PHOTOGRAPH
YOU TAKE.

CENTER FOR
ASTROPHYSICS
HARVARD COLLEGE
OBSERVATORY
SMITHSONIAN ASTROPHYSICAL
OBSERVATORY

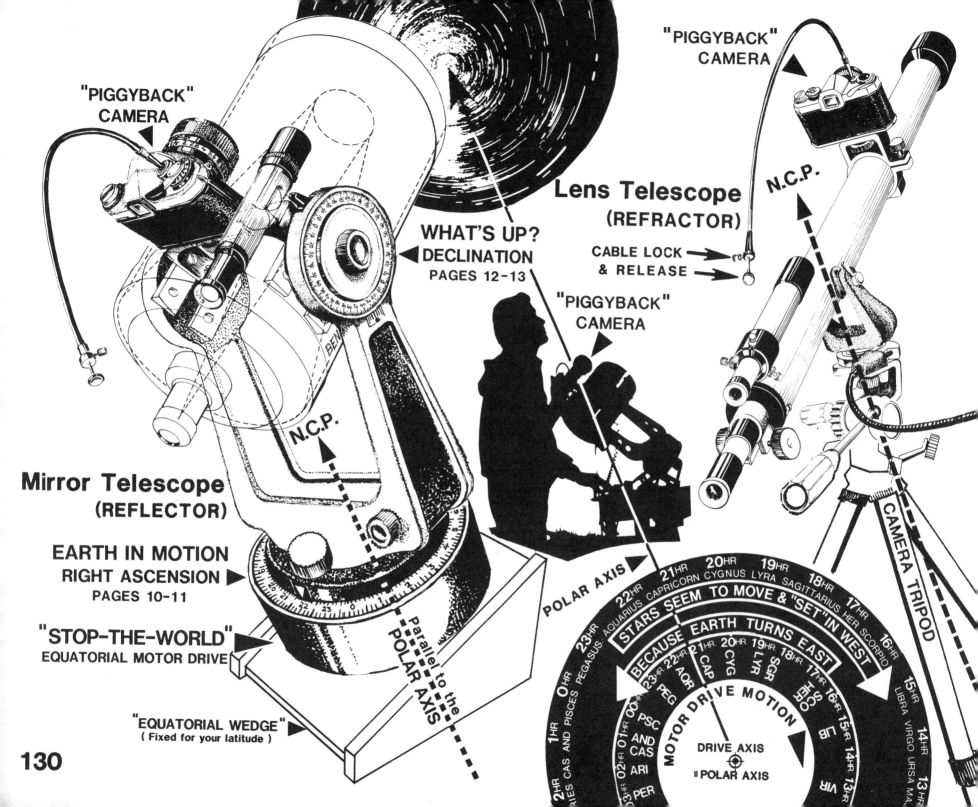

"PIGGYBACK" CAMERA

"PIGGYBACK" CAMERA

Lens Telescope
(REFRACTOR)

CABLE LOCK →
& RELEASE →

N.C.P.

WHAT'S UP?
DECLINATION
PAGES 12-13

"PIGGYBACK" CAMERA

N.C.P.

Mirror Telescope
(REFLECTOR)

EARTH IN MOTION
RIGHT ASCENSION ▶
PAGES 10-11

"STOP-THE-WORLD"
EQUATORIAL MOTOR DRIVE

"EQUATORIAL WEDGE"
(Fixed for your latitude)

Parallel to the POLAR AXIS

CAMERA TRIPOD

POLAR AXIS ▶

STARS SEEM TO MOVE & "SET" IN WEST

BECAUSE EARTH TURNS EAST

MOTOR DRIVE MOTION

DRIVE AXIS
= POLAR AXIS

20HR 19HR 18HR 17HR
21HR SAGITTARIUS HER SCORPIO
22HR CAPRICORN CYGNUS LYRA 16HR
23HR PEGASUS. AQUARIUS 15HR LIBRA VIRGO URSA MA
0HR AND PISCES 14HR
1HR ARIES CAS 13HR
2HR
ARIES CAS AND PISCES
CAP CYG SGR
AQR LYR SCO
PEG 18HR HER
PSC 19HR 16HR LIB
AND 20HR 17HR 15HR
CAS 21HR VIR
ARI 22HR 14HR
PER 23HR 13HR

130

HALF A TELESCOPE

Once we know the direction in which the earth rotates, it stands to reason that any drive mechanism designed to turn cameras or telescopes in the opposite direction and at the same slow rate of one revolution in twenty-four hours will nullify this apparent motion and keep stars in one place while they are being photographed or observed. Such a mechanism can be built or be purchased ready-made. So-called "equatorial drives" form part of almost all better telescope systems today. They quite literally "stop the world" and permit you to take long exposures where all stars form crisp round-point images, and where even faint stars are given the time to let their light accumulate on just one spot of the film. Then you will be able to photograph, print and project quality color starframes to match the twenty-four pairs in this book.

For a start, you may wish to acquire only the motorized half of a sizable telescope, just so that you can attach a small support bracket to it to move your camera in right ascension and swing it in declination. Such a combination will even support longer camera telelenses, which are telescopes in themselves. It will open the sky to you and allow you to collect the colorful starlight which telescope catalogues so temptingly depict, but which the eye alone can never see. Your every photograph puts you on the threshold of real astronomical discovery as soon as you start blinking with a Problicom (pages 132-33).

If you have your eye and pocketbook set on a complete telescope system, and there will be no stopping you if you do, then spend the tiny extra sum and purchase a "camera clamp." This is the most important and least expensive accessory. It enables you immediately to mount your camera on top of the telescope in the "piggyback" mode.

Do not — repeat, *do not* — attach your camera to the eyepiece position until you have familiarized yourself with simple motorized 50mm-lens photography. Telescopes are, in effect, extremely long telelenses, but, because they are relatively "slow" (f/10 or slower), very long exposures are needed to capture any but the brightest astronomical images. At telescopic magnifications every error, any misalignment, is also blown up

DRIVE TELESCOPE WITH MANUAL LINKAGE WHILE KEEPING A STAR CENTERED IN EYEPIECE.

LUNAR EXPOSURES THROUGH TELESCOPES CAN RANGE FROM 1/250 TO 5 SECONDS. KEEP ACCURATE NOTES, TEACH YOURSELF.

enormously. Just as immediate success can virtually be guaranteed with 50mm piggyback astrophotography, failure is almost predictable, at first, with through-the-telescope work. It is advisable to increase one's experience along with optical focal length.

When we speak of alignment, it refers to the important fact that for any motorized drive system to work, its axis must be pointed fairly accurately at the North Celestial Pole (N.C.P., page 36). For 50mm piggyback photography the "polar alignment instructions" which are packed with every drive system or telescope are quite adequate when followed. Refer to the previous chapter concerning the doubling of the lengths of exposures. Now you can start with thirty seconds, then go to a minute. Double to two, then four minutes. Next go to eight minutes. Under good dark skies a quarter-hour photo will produce what you find in the left-hand pages of this book.

As soon as photography *through* telescopes is attempted, combining high magnification with long exposure times, it becomes imperative that the axis of your telescope drive be *exactly* parallel to the polar axis on which our earth turns. So, if you must shoot through your telescope immediately, try lunar photography. Here is an ideally large object which even changes in brightness, but requires only short exposures. By all means, shoot for the moon. Be sure you take notes as you go.

THE BLINKING PRINCIPLE:

To obtain a "Blinking" effect, flip this corner starfield quickly back and forth over the one below (on page 133). Find a nova, a variable star, a comet and an asteroid.

REFERENCE SLIDE: SLIDE TAKEN LAST WEEK LAST MONTH OR EVEN LAST YEAR.

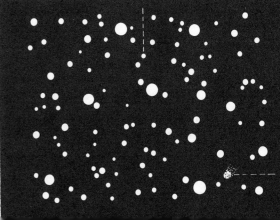

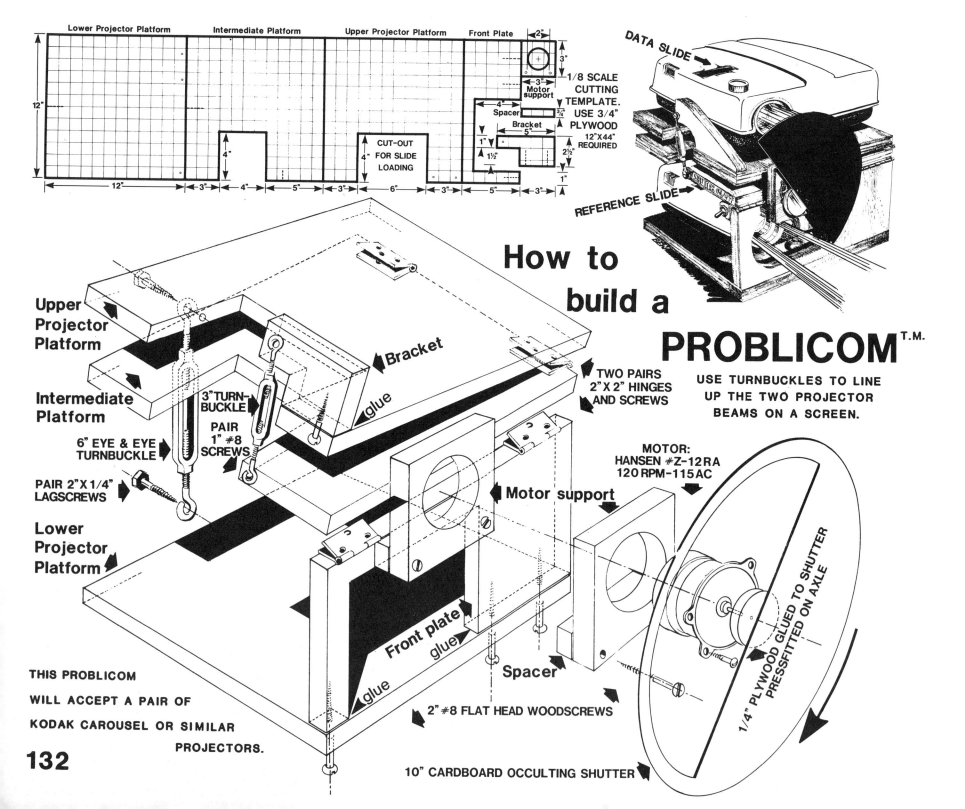

Lower Projector Platform

Intermediate Platform

Upper Projector Platform

Front Plate

2"

3"

Motor support

3"

4" Spacer 3/4"

Bracket 5"

1" 1½"

2½"

1"

12"

4"

CUT-OUT FOR SLIDE LOADING

4"

1/8 SCALE CUTTING TEMPLATE. USE 3/4" PLYWOOD 12"X44" REQUIRED

12"   3"   4"   5"   3"   6"   3"   5"   3"

DATA SLIDE

REFERENCE SLIDE

How to build a

PROBLICOM™

USE TURNBUCKLES TO LINE UP THE TWO PROJECTOR BEAMS ON A SCREEN.

Upper Projector Platform

Bracket

glue

Intermediate Platform

3" TURN-BUCKLE

PAIR 1" #8 SCREWS

6" EYE & EYE TURNBUCKLE

PAIR 2" X 1/4" LAGSCREWS

Lower Projector Platform

TWO PAIRS 2" X 2" HINGES AND SCREWS

MOTOR: HANSEN #Z-12RA 120 RPM-115AC

Motor support

Front plate glue

Spacer

glue

1/4" PLYWOOD GLUED TO SHUTTER PRESSFITTED ON AXLE

THIS PROBLICOM WILL ACCEPT A PAIR OF KODAK CAROUSEL OR SIMILAR PROJECTORS.

2" #8 FLAT HEAD WOODSCREWS

10" CARDBOARD OCCULTING SHUTTER

"DISCOVERY MACHINE"/PROBLICOM

PLANET NEPTUNE

ASTEROID CERES

There is a special serendipity factor (the gift of finding things by chance or where one least expects them) in astrophotography which can turn seemingly unimportant star photos into potential history-makers. Any image, however badly exposed, may be a discovery or pre-discovery photograph. It has happened time and time again.

Where last month there may have been 1,000 stars in a picture of a certain region of the sky, last night there may have been 1,001. You will never know unless you have a Problicom "discovery machine" to help you find the new arrival, be it a nova, a variable star or even a comet. By the alternate superimposing of two projected slide images onto a screen, pictures taken of the same starfield but on different nights allow us to make valuable astronomical discoveries. This simple method is called "blink comparison."

A rotating occulting shutter turning in front of the stacked projector lenses presents the two pictures in rapidly alternating succession to the viewer(s). The two images will be seen as one, with stars appearing static and in their place — all except newcomers or objects which have moved their position or changed in brightness. Those will instantly attract attention to themselves by "blinking" on and off or by jumping back and forth on the screen at the rate at which the small motor drives the occulting shutter to interrupt the two projection beams.

Blinking is an elegant process. It is a logical byproduct of modern photography. Professional blink comparators are complex optical devices with price tags in five or six figures. The author has developed the Problicom concept, which allows even amateurs to perform meaningful work with a relatively inexpensive home-built system as illustrated. No familiarity with the stars is needed.

You can stake out your own starframe areas and patrol them with simple photography. Your first pair of slides from the region of the Zodiac can help you find Neptune or Uranus and to chart their motions along with those of Saturn or Mars (pages 14-15). Asteroids will signal their changing positions.

Variable stars will reveal their varying magnitudes and allow you to start observing and recording their changing luminosities photographically (pages 121-22). A casually taken roll of star photos may yield a precious nova and bring you the pride of genuine scientific achievement (page 83). In a systematic and disciplined predawn "haystack" survey (pages 140-41) you can discover a comet which will bear your name for all time. Problicom puts you on the threshold to discovery. Your research can bring you scientific immortality.

HOW BLINKING WORKS:

For a simple demonstration of how PROBLICOM works, rapidly flip the starfield on page 131 back and forth over the one shown here. Blink-comparison will soon explain itself.

VARIABLE X CYG

NOVA CYGNI 1978

COMET WEST

DATA SLIDE: YOUR MOST RECENT SLIDE OF THE SAME AREA AS ON PAGE 131.

CITY SKIES/COUNTRY SKIES

(SEE PAGES 80-81)

ε   α

δ   ζ

γ   β

M 57

There is no such thing as a bad night sky.

Perhaps you live in a city apartment, and from your location you can see only a limited patch of the sky. Do not give up. As long as any part of the celestial sphere can be observed, the moving earth will bring you a year-round panorama. It may have to be viewed out a window, but there will be a new view each month.

Determine the orientation of your view. A street map will help. North is always at the top, east to the right, west to the left. A janitor's key and access to a roof may be your passport to heaven. From such a location you will be able to see at least part of the ecliptic where the bright planets travel. You should have several starframes filled with constellations. Bright stars, once you know their names, will be your entry ticket to the celestial carousel.

Still, city dwellers are often limited in their ability to see the lower (southern) declinations. Structures may block their view, and invariably city lights will hinder them. Light-pollution may improve after midnight, when a surprising number of lights are extinguished. Make test exposures at 10 P.M., at midnight, even at 2 A.M., to find the best observing times.

A bus ride to the center of a safe park or dark playing field (ideally on higher ground) will guarantee better viewing. For maximum improvement, each mile away from the center of even a small town will be well rewarded. Go south, if you can, to put the city glow behind you beneath the North Star. Then you will see as much of the celestial melon slices as possible while they culminate at the times as listed.

Explore different sites, and do not rely on just your memory for sky conditions. Employ scientific methods. Use a camera. Load a roll of ISO 400 *black-and-white* film and shoot a 1-2-4-8-15-30-60-second spread. Before shooting from cities, buy an inexpensive red filter to place over your lens to cut down the glow of the city lights. (Ask for a #25 Wratten filter in any camera store). Shoot the above salvo again and add a two-minute exposure. Take notes as you go — please take good notes. Then review and compare your negatives. The differences should surprise you.

"Empirical learning" means getting firsthand information for oneself, from actually experimenting. It is the best way to teach oneself. Read books on camera filters if you must, but be forewarned: they may scare you with graphs and formulas. It's the red filter with black-and-white films you are after when you shoot from the city. If you doubt it, smooth out a red cellophane candy wrapper to test the concept. Yes, you can collect starlight even under city skies, and you can blink black-and-white film negatives after mounting them, just like color slides, in little cardboard mounts sold for this purpose. Developing black-and-white film can be easy and quick. Ask for details in a photo-supply store.

Your first trip to a truly dark observing site will be a revelation. It will be love at first sight. Invite children, take a loved one. Nothing can equal the splendor of a velvet-black firmament with myriads of stars. You may find much more than the starry heavens — you may discover yourself.

Afterward, any glimpse of the night sky can prove that you are not living in a confining apartment but are journeying on a splendid planet around a star, a solar-system sailor in a magnificent galaxy — citizen of a vast universe.

LYRA CLOSE-UPS FROM CITY:
135MM TELE-LENS PHOTO ···· ······ EQUALS BINOCULAR FIELD ·····OF VIEW.

FILM: EKTACHROME ISO 400
EXPOSURE: 4 MINUTES
APERTURE: F 2.2

Photography from brighter locations will produce amazing results with very short exposures but pollution sky-glow may soon begin to fog a starfield, limiting photography and turning backgrounds into various shades of gray.

STARS BEYOND 6TH MAGNITUDE

FROM COUNTRY:

EXPOSURE: 20 MINUTES
APERTURE: F 4

The darker your astrophotography site, the longer exposures can be. Good contrast can be maintained, permitting penetration to much fainter magnitudes. Short exposures from dark areas will also be of better color and detail.

STARS TO 10TH MAGNITUDE   **NOTE M57**

RED COVER

ADAPTING TO THE DARK

Red cellophane candy wrappers are very useful tools in astronomy. For the following application any red translucent material will do, even some crimson tissue paper or cloth. Stretch it over the bulb of your flashlight with tape or a rubber band to cut down on the glare of white light. To read starframe charts at night, to make notes or to work a camera, always use such a red light source. It will help your eyes maintain their adaptation to the dark.

It takes our eyes a fair amount of time to get accustomed to the dark, as anyone knows who has ever tried to find a seat in a dark theater. We have learned from experience that after a few minutes our eyes adjust to the conditions and we are able to see what was hidden to us before.

In astronomy this wonderful natural gift is of greatest importance. Time and again a variety of conditions are blamed for the inability to see stars or a faint comet, when, in effect, impatience was the cause. It is simply impossible to go from a bright room into the night and to make observations (unless they involve the moon). At least ten or even fifteen minutes must be allowed for the eyes to become accustomed to the dark. City dwellers most of all should rest their eyes awhile in near-darkness. They will then be able to see stars of fainter magnitude than they considered possible.

Once this state of sensitivity has been achieved, even one brief look at a bright light or a television screen can spoil it, requiring a new period of readjustment for the eye. That is why the dimmed red flashlight is so important. Red light does not disturb the eye as much as white does.

In old cameras the aperture controlling the size of the lens opening was called the "iris" because it performs exactly the same function as the iris in the eye. Both regulate the amount of light which can pass through. Under bright conditions the pupils in our eyes protectively contract, just as we have to "stop down" lenses to avoid overexposures in bright sunlight.

At night, when our eyes are fully adapted to the dark, pupils are at their largest, admitting the maximum amount of light through the iris. At night, our camera lenses too must be set at their widest possible opening. We want to achieve maximum aperture for best night-seeing, both for our camera and for our eyes.

EYE   **LENS**

IN DAYLIGHT   **STOPPED DOWN**

IN DARKNESS   **WIDE OPEN**

New

Waning*

Full

CRATERS: COPERNICUS   TYCHO (W.RAYS)   APPENINES (MOUNTAINS)

PLATO

EARTH–MOON DISTANCE 237,000 MILES

EARTH–SUN DISTANCE 93,000,000 MILES

ECLIPTIC

SUN

CLAVIUS

PHOTOGRAPHY THROUGH AN 8" DIA. TELESCOPE. FILM: KODACHROME ISO 25. APERTURE NON-ADJUSTABLE.

The moon, our nearest celestial neighbor, offers a fine target for visual exploration with binoculars or fast photography through telescopes. Some of the most prominent lunar features are listed here. Sky atlases give the names of hundreds more.

As the moon orbits the turning earth every lunar month it shows us only one side. Spacecraft are needed to explore the hemisphere which faces away from us. Man first set foot on the moon in July 1969. "One giant leap for mankind."

The line separating the sunny and the shady areas is called the "terminator." Craters are best seen on or near this edge, where sunlight and shadows sharply define the lunar details. Moon surfaces when exposed to the sun reach temperatures of +95°C (+203°F). In the shade they drop to -150°C (-238°F). The moon has a north and a south pole. The photographs shown here have "north up." This is how we see it in the sky or through binoculars. For telescopes which reverse images you will have to turn the book upside down to identify the various landmarks.

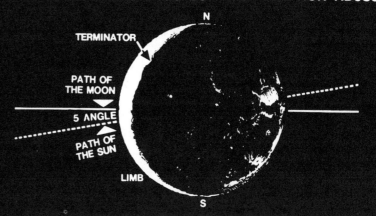

N

TERMINATOR

PATH OF THE MOON

5 ANGLE

PATH OF THE SUN

LIMB

S

The familiar phases of the moon depend on its position in relation to the sun. To understand these phases, just think of where the sun is at any given time to illuminate the moon from the right, the left or, at full moon, the front. At new moon, the lunar disc is lost to us in the glare of the sun, which lies beyond the earthbound moon.

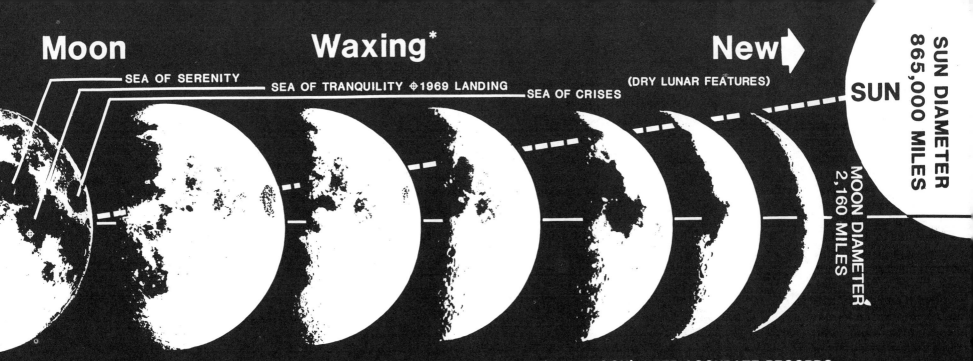

Moon   Waxing*   New ▶

SEA OF SERENITY — SEA OF TRANQUILITY ⊕ 1969 LANDING — SEA OF CRISES — (DRY LUNAR FEATURES)

SUN

SUN DIAMETER 865,000 MILES

MOON DIAMETER 2,160 MILES

EXPOSURES: FROM 3 SECONDS (PARTIAL PHASES) TO 1/15 SECOND (FULL MOON). KEEP ACCURATE RECORDS.

The moon is much smaller than the earth, but close to us. The sun is enormously larger (page 14) but far, far away. This gives the appearance that they are the same size. Nothing could be further from the truth.

The orbit of the moon around the earth and the orbit of the earth around the sun form planes which are very slightly inclined toward each other. This inclination of five degrees means that the moon may travel anywhere from five degrees above to five degrees below the ecliptic — well within the Band of the Zodiac.

If this were not so, then every new moon would automatically create a total solar eclipse. Lunar eclipses would come with every full moon.

Actually, the correct alignment of sun, earth and moon occurs rarely, making eclipses spectacular, special events.

*Waxing or waning?

To plan a productive observing program it is useful to know the moonphases in advance. The zooming/abating jingle based on the directions of the curves of the script letters a and z may help to show and tell where the moon is going.

Before and during first quarter, the moon rises progressively later in daytime to set ever closer to midnight.

Approaching full moon and beyond, the bright disc rises between dusk and first darkness, to set with daylight.

Last quarter has the moon rising ever nearer midnight affording increasingly long dark evenings for starwatching.

Near new moon, sun and lunar disc rise and set together offering longest hours of darkness for best astrophotography.

Check your newspaper for all-dark new moon nights.

Last /4	New	First/4	Full
Jan. 29	Feb. 6	Feb. 12	Feb. 21

AN EASY WAY TO REMEMBER

LUNAR A to Z

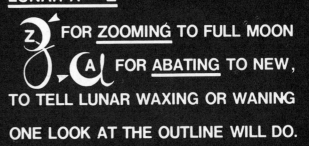

z FOR ZOOMING TO FULL MOON

a FOR ABATING TO NEW,

TO TELL LUNAR WAXING OR WANING

ONE LOOK AT THE OUTLINE WILL DO.

BEN

137

SUN

SOLAR ECLIPSE AFRICA/INDIA

SUN/MOON/EARTH DISTANCES AND DIAMETERS ON PAGES 136–37

DIAMOND RING EFFECT

NEAR END OF ECLIPSE

KENYA, 16 FEBRUARY 1980

SUN

During a lunar eclipse, the moon is in the shadow of the earth. In a solar eclipse the light of the sun is blocked by the moon.

When the moon crosses the plane of the ecliptic, the stage may be set for a lunar or solar eclipse. As a general rule, lunar eclipses are visible from relatively large areas on earth (at night only) while the small moon dwells in the shadow of the larger earth. For the same reason lunar eclipses are likely to occur more frequently than solar.

OF THE MOON⋯⋯

ECLIPSES

Solar eclipses come in two varieties, depending on the changing distances and apparent diameters of the sun in relation to the moon. When a sun-moon-earth alignment comes at a time when the moon is somewhat farther from us than usual, the diameter of the sun appears to us to be larger than that of the occulting moon, and a dazzling halo-like ring of the sun remains visible even as the moon passes in front of it. During such an "annular" solar eclipse, sunlight may be noticeably dimmer for a few moments, but it does not get truly dark.

A total solar eclipse occurs when the moon is nearer the earth and when the lunar disc completely hides the sun. This is the rarest of eclipses. It is visible only from the narrow path along which the moon's shadow races over parts of the earth. Darkness in any one place may range in duration from seconds to minutes. The corridor of darkness may be some 50 to 100 miles wide. Among the grand spectacles of nature, a total eclipse — when day suddenly turns to night — is the most awesome and beautiful.

SOLAR ECLIPSE U.S./CANADA, AT 32,000 FT. BY JET, 26 FEB. 1978

SOLAR ECLIPSE SURABAYA, INDONESIA, 11 JUNE 1983

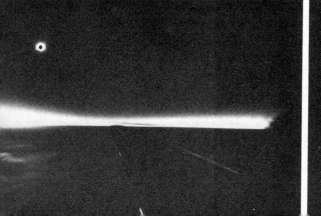

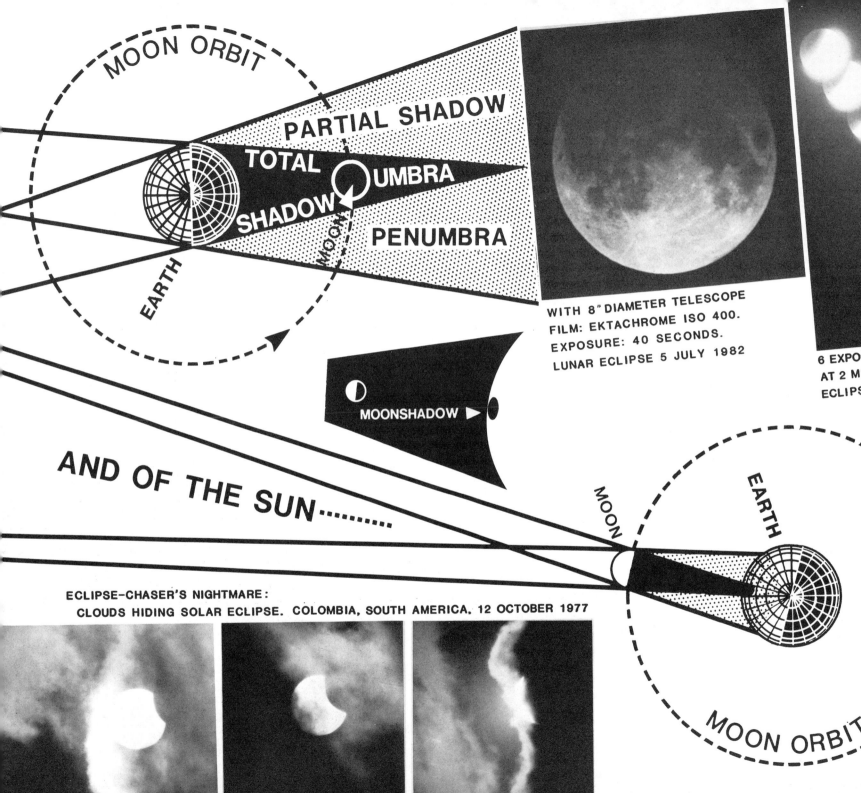

MOON ORBIT

PARTIAL SHADOW

TOTAL

SHADOW UMBRA

PENUMBRA

MOON

EARTH

WITH 8" DIAMETER TELESCOPE
FILM: EKTACHROME ISO 400.
EXPOSURE: 40 SECONDS.
LUNAR ECLIPSE 5 JULY 1982

6 EXPOSURES 1/4 SEC.
AT 2 MIN. INTERVALS
ECLIPSE 29 NOV. 1974

MOONSHADOW ▶

AND OF THE SUN..........

MOON

EARTH

ECLIPSE-CHASER'S NIGHTMARE:
CLOUDS HIDING SOLAR ECLIPSE. COLOMBIA, SOUTH AMERICA, 12 OCTOBER 1977

MOON ORBIT

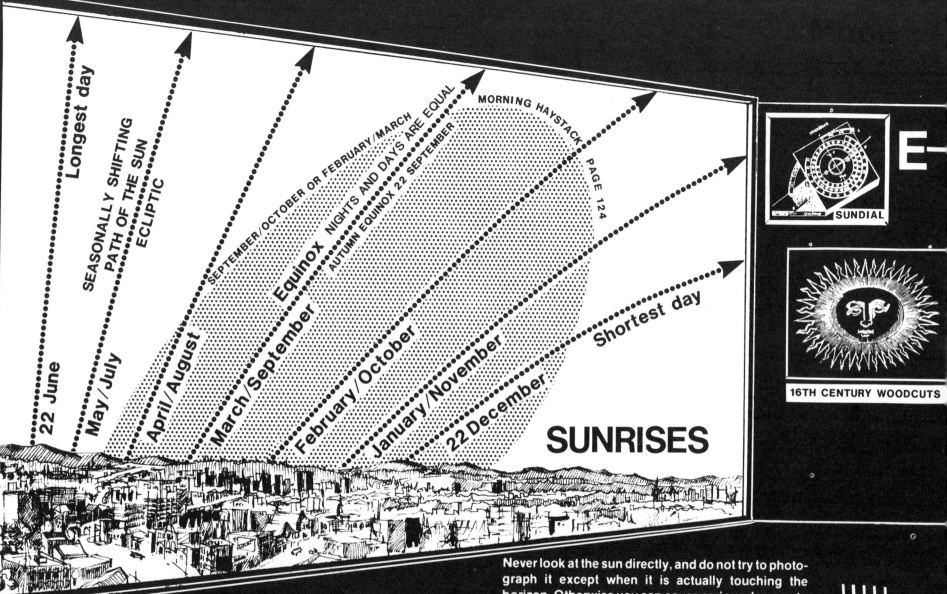

Longest day

22 June

May/July

SEASONALLY SHIFTING
PATH OF THE SUN
ECLIPTIC

April/August

SEPTEMBER/OCTOBER OR FEBRUARY/MARCH

Equinox NIGHTS AND DAYS ARE EQUAL

AUTUMN EQUINOX 22 SEPTEMBER

MORNING HAYSTACK PAGE 124

March/September

February/October

January/November

22 December

Shortest day

SUNRISES

E-

SUNDIAL

16TH CENTURY WOODCUTS

SUN
BEAM

HOT
SUMMER

SUNSETS/SUNRISES

The dawning of a new day and the glow of sundown share poetic beauty. If one keeps notes to indicate the positions of the risings and the settings of the sun, together with a midday position record, it is easy to follow the changing angles of the ecliptic in relation

Never look at the sun directly, and do not try to photograph it except when it is actually touching the horizon. Otherwise you can cause serious damage to your eyes or your photographic equipment.

Shadows cast by the sun will serve to indicate its angle and position on any clear day. For you the sun may first appear over a group of trees and it may shed its last rays as it disappears behind some building. Chalk-mark some shadow lines. They can all be traced back to the dazzling solar disc. They will confirm the view from our imaginary penthouse room

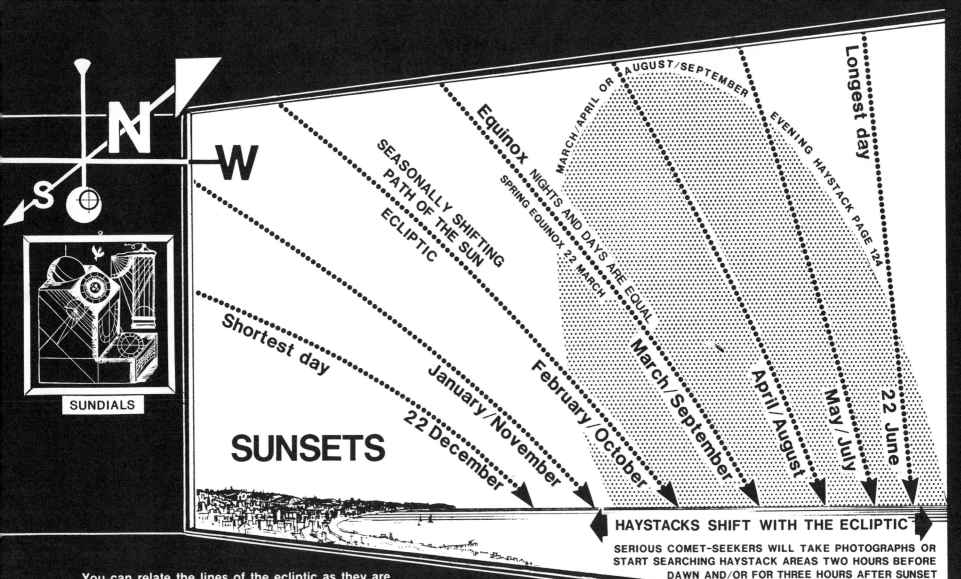

N

W

S

SUNDIALS

SEASONALLY SHIFTING
PATH OF THE SUN
ECLIPTIC

Equinox NIGHTS AND DAYS ARE EQUAL
SPRING EQUINOX 22 MARCH

MARCH/APRIL OR AUGUST/SEPTEMBER

EVENING HAYSTACK PAGE 124

Longest day

Shortest day

January/November

February/October

March/September

April/August

May/July

22 June

22 December

SUNSETS

HAYSTACKS SHIFT WITH THE ECLIPTIC

SERIOUS COMET-SEEKERS WILL TAKE PHOTOGRAPHS OR
START SEARCHING HAYSTACK AREAS TWO HOURS BEFORE
DAWN AND/OR FOR THREE HOURS AFTER SUNSET
(SEE PAGE 124)

You can relate the lines of the ecliptic as they are drawn in all the starcharts of Zodiac constellations to the curved path which the sun travels. The moon, when seen in daytime, is never far from the sky lane of the sun. At night the moon and the planets confirm the changing angle of the celestial corridor.

Morning and evening "haystacks" are shown. This is where comet seekers search the skies well before dawn or after sunset — twilight and on into the night (page 124). Comets may be discovered anywhere in the sky, but chances are improved when the shown areas are photographed and blinked with regularity.

No wonder summers are hot when the sun beats straight down. In winter the rays come in at an angle and spread out over larger areas. When the ecliptic is "low" for us in the Northern Hemisphere, it is "high" for our neighbors south of the equator. That is when the Southern Hemisphere enjoys its long days of summer.

High or low, the sun is always "out" as we leisurely sail and spin around it. Sunrise, sunset, sunrise, sunset ...

SUN BEAM

COLD
WINTER

141

RESOURCES

BIBLIOGRAPHY AND RECOMMENDED READING

Abell, George O., *Exploration of the Universe,* 4th edition. Philadelphia: Saunders College Publishing, 1982.

Burnham, Robert, Jr., *Burnham's Celestial Handbook,* 3 volumes, New York: Dover Publications, 1978.

Kukarkin, B.V. and others, *Catalogues of Variable Stars,* 3 volumes and supplement. Moscow: U.S.S.R. National Academy of Sciences, 1976.

Kunitzsch, Paul, *Arabische Sternnamen in Europa.* Wiesbaden, Germany: Otto Harrassowitz, 1959.

Mallas, J.H. and E. Kreimer, *The Messier Album.* Cambridge, MA: Sky Publishing Corporation, 1978.

Menzel, D.H., and J.M. Pasachoff, *A Field Guide to the Stars and Planets,* 2nd edition. Boston: Houghton Mifflin Co., 1983.

Morrison, Philip and Phyllis, *Powers of Ten.* San Francisco: W.H. Freeman & Co., 1982.

Pasachoff, Jay M., *Astronomy from the Earth to the Universe.* Philadelphia: W.B. Saunders Co., 1979.

Vehrenberg, Hans, *Atlas of Deep Sky Splendors.* Cambridge, MA: Sky Publishing Corporation, 1978.

TELESCOPES AND RELATED EQUIPMENT

Bushnell/Bausch and Lomb
2828 East Foothill Boulevard, Pasadena, CA 91107

Celestron International
2835 Columbia Street, Torrance, CA 90503

Coulter Optical
P.O. Box K, Idyllwild, CA 92349

Edmund Scientific
101 E. Gloucester Pike, Barrington, NJ 08007

Meade Instruments
1675 Toronto Way, Costa Mesa, CA 92626

Orion Telescopes
P.O. Box 1158-T, Santa Cruz, CA 95061

Questar Corporation
P.O. Box C, New Hope, PA 18938

Roger W. Tuthill
11 Tanglewood Lane, Mountainside, NJ 07092

MAGAZINES

Astronomy/Deep Sky
Astromedia Corporation,
P.O. Box 92788, Milwaukee, WI 53202

Griffith Observer
Griffith Observatory,
2800 East Observatory Road, Los Angeles, CA 90027

Mercury
Astronomical Society of the Pacific,
1290 24th Avenue, San Francisco, CA 94122

Sky and Telescope
Sky Publishing Corporation,
49 Bay State Road, Cambridge, MA 02238

Telescope Making
Astromedia Corporation,
P.O. Box 92788, Milwaukee, WI 53202

STAR ATLASES

Norton's Star Atlas
Sky Publishing Corporation,
49 Bay State Road, Cambridge, MA 02238

Tirion Atlas 2000.0
Sky Publishing Corporation,
49 Bay State Road, Cambridge, MA 02238

AAVSO Variable Star Atlas
American Association of Variable Star Observers,
187 Concord Ave., Cambridge, MA 02138

POSTERS, SLIDES, PHOTOGRAPHS

Hansen Planetarium Publications
15 South State Street, Salt Lake City, UT 84111

LASER VIDEO DISCS

Video Vision Associates Ltd.,
7 Waverly Place, Madison NJ 07940

SKY CALENDAR (Monthly)

Abrams Planetarium,
Michigan State University, East Lansing, MI 48824

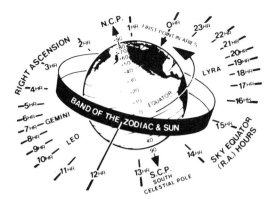

GLOSSARY

A.A.V.S.O. American Association of Variable Star Observers.

APERTURE. Effective light-gathering diameter such as the opening of a lens.

ASTEROID. A minor planet.

CELESTIAL EQUATOR. A great circle on the celestial sphere beyond and in the same plane as the equator of the earth.

COMET. A body of ices, rock and dusty matter.

CONSTELLATION. A grouping of stars named for mythical figures, animals or objects.

DAYLIGHT SAVING TIME. Time more advanced by one hour than standard time.

DECLINATION (Dec). The angle north or south of the equator to an object measured from the twenty-four hour circle of the celestial equator.

ECLIPSE. The partial or complete cutting off of the light of a body by another passing in front of it.

ECLIPTIC. The apparent sky path of the sun.

EQUATORIAL DRIVE. A motorized mounting with an axis parallel to the earth's axis whose motion compensates for the rotation of our planet.

EQUINOX. One of the two intersections of the ecliptic and the celestial equator.

FLATFORM. The seemingly flat observing platform on the surface of the round earth.

GALACTIC CLUSTER. An "open" cluster of stars located within the disk of our galaxy.

GALAXY. A vast assemblage of millions to hundreds of thousands of millions of stars.

GLOBULAR CLUSTER. Ball-shaped groupings of stars above and below the galactic plane.

HAYSTACK. Horizon zones of haystack shape centered on the ecliptic, where comets can be found before sunrise and/or after sunset.

LATITUDE. A north-south coordinate on the surface of the earth.

LIGHT-YEAR. A measure of distance, not time. The distance light travels in one year.

MAGNITUDE. A scale of measurement for the brightness of stars and other sky objects.

METEOR. A body of rock or metal heated to incandescence when it enters the atmosphere of the earth. Also called a "shooting star."

METEORITE. Surviving part of a meteor which strikes the earth after a fiery descent.

METEOR SHOWER. A display of many meteors radiating from a common point in the sky.

MILKY WAY. A wide band of light stretching around the celestial sphere caused by the light of myriads of faint stars.

NEBULA. Cloud of interstellar gas or dust glowing from or reflecting nearby starlight.

NOVA. A star that suddenly undergoes an outburst of radiant energy and increases its luminosity by hundreds or thousands of times.

OPEN CLUSTER. See Galactic Cluster.

ORBIT. The path of a body in its revolution about another body or center of gravity.

PLANET. One of nine large spherical objects revolving about the sun and shining by reflected light.

PLANETARY NEBULA. A shell of gas blasted from a dying star.

PRECESSION. A cone-shaped motion of the axis of the earth's rotation caused mostly by the sun and the moon.

PROBLICOM. A PROjection BLInk COMparator to present two different photographs of the same region of the sky for easy comparison.

RADIANT. The point on the celestial sphere from which meteors of a given shower seem to originate.

RETROGRADE MOTION. Apparent "backward" (or westward) drift of a planet on the celestial sphere with respect to the stars, resulting from the motion of the earth.

RIGHT ASCENSION (R.A.) A coordinate for finding the east-west positions of celestial bodies, measuring eastward along the celestial equator from the spring equinox to the point below or above the body in question.

SCALE. The linear distance in a photograph corresponding to a particular angular distance in the sky, i.e., so many fractions of an inch per arc-minute or per degree.

SERENDIPITY. The gift of finding valuable or useful things or data not sought for.

SPORADIC METEOR. A meteor (shooting star) which does not belong to a meteor shower.

STARFRAME. A sky area defined by a specific outline (such as a bent coathanger) to contain the principal stars of a constellation for easy repeated finding and study.

STAR PARTY. A gathering of amateur star-lovers on new-moon weekend nights to view the skies, take photographs, compare equipment and share astronomical information.

SUPERNOVA. A stellar cataclysm in which a star explodes, briefly increasing its luminosity by hundreds of thousands to hundreds of millions of times.

TERMINATOR. The dividing line between the sunlit and shadowed portions of the moon or a planet.

VARIABLE STAR. A star whose brightness varies.

ZENITH. The point directly overhead, the direction opposite to that of a plumb bob.

ZODIAC. A band around the sky which has the ecliptic in its center and in which the sun, moon and planets move.

INDEX